The Michael C. .khov Handbook

'Petit's words go right to the heart of Chekhov's technique . . . Anyone looking for a key to understanding more about Michael Chekhov's technique will devour it.' – *Jessica Cerullo, Michael Chekhov Association, NYC*

The Michael Chekhov technique is today seen as one of the most influential and inspiring methods of actor training in existence. In *The Michael Chekhov Handbook*, Lenard Petit draws on twenty years of teaching experience to unlock and illuminate this often complex technique for those studying, working with or encountering it for the first time.

Petit uses four sections to help readers approach the technique:

- the aims of the technique – outlining the real aims of the actor
- the principles – acting with energy, imagination and creative power
- the tools – the actor's use of the body and sensation
- the application – bringing the technique into practice.

The Michael Chekhov Handbook's explanations and exercises will provide readers with the essential tools they need to put the rewarding principles of this technique into use.

Lenard Petit is the Artistic Director of The Michael Chekhov Acting Studio in New York City. He teaches Chekhov technique in the MFA and BFA Acting programmes at Rutgers University. He was a contributor and co-creator of the DVD *Master Classes in the Michael Chekhov Technique*, published by Routled~~

The Michael Chekhov Handbook

For the Actor

Lenard Petit

Routledge
Taylor & Francis Group

LONDON AND NEW YORK

First published 2010
by Routledge
2 Park Square, Milton Park, Abingdon, Oxon, OX14 4RN

Simultaneously published in the USA and Canada
by Routledge
270 Madison Ave, New York, NY 10016

*Routledge is an imprint of the Taylor & Francis Group,
an informa business*

© 2010 Lenard Petit

Typeset in Joanna and ScalaSans by
Florence Production Ltd, Stoodleigh, Devon

Printed and bound in Great Britain by
TJ International Ltd, Padstow, Cornwall

British Library Cataloguing in Publication Data
A catalogue record for this book is available from the British Library

Library of Congress Cataloging in Publication Data
Petit, Lenard.
  The Michael Chekhov handbook: for the actor/By Lenard Petit.
    p. cm
  1. Acting.   2. Chekhov, Michael, 1891–1955.   3. Title.
  PN2061.P475 2009
  792.02′8—dc22   2009007095

ISBN10: 0–415–49671–3 (hbk)
ISBN10: 0–415–49672–1 (pbk)
ISBN10: 0–208–87230–4 (ebk)

ISBN13: 978–0–415–49671–1 (hbk)
ISBN13: 978–0–415–49672–8 (pbk)
ISBN13: 978–0–203–87230–7 (ebk)

This book is dedicated to the
two sustaining loves of my life,
Meg and Luke Pantera/Petit

CONTENTS

ACKNOWLEDGEMENTS

I would like to thank Talia Rodgers and Ben Piggot at Routledge for taking this book idea seriously and guiding me through the process of its acceptance and release.

I would like to thank all my teachers who pointed me in the directions I have gone in. This book could not have happened without them. None of them are still with us, so I'll keep their names in my heart.

Thanks to PAJ Books for permission to include some words of Michael Chekhov from their publication, *Lessons for the Professional Actor*.

I am very grateful to Rutgers University for giving me the playground to develop this material over a number of years; with great support from Carol Thompson, Barbara Marchant, Kevin Kittle, Deborah Hedwall and Heather Rasche. I should also like to thank Maggie Flanigan and the actors at her studio in New York City for allowing me to include their responses in this book.

A special thanks is reserved for Bethany Caputo and Judith Bradshaw for transcribing numerous workshops and including invaluable observations about the results of teaching this work. Also to James Luse, Janet Morrison and Mel Shrawder for reading the manuscript and assisting me in its presentation. To Edward Marritz for his photographic eye.

I need to thank my colleagues at MICHA: Joanna Merlin, Jessica Cerullo, Marjolein Baars, Ted Pugh, Fern Sloan, David Zinder, Sarah Kane and Ragnar Freidank for their confidence and enthusiasm for my ideas and work.

I also want to thank my students, actors who continually inspire me to create new things for them to play with, but especially Scott Miller, Sal Cacciato, Jessie Greene, Tyree Giroux, Ben Bauman, John Rawlinson, Brian Parrish, Jonathan Day, Bryan

Cohen, Jessica Savage, Glenn Cruz, Caron Levis, Oliver Martin and Juliette Bennett, for showing up all the time and digging into the work.

The biggest and most profound thanks goes to Michael Chekhov, whose genius continues to vibrate all around me.

INTRODUCTION

If I could have, I would have asked Michael Chekhov: What do you believe is the most important factor in your method? Instead, I asked my teacher, Deirdre Hurst du Prey, Chekhov's student and secretary for 20 years. 'That would be Truth', she answered. This is a big idea, but simple enough. The difficulty is in keeping it simple. Once it gets complicated the truth becomes more and more elusive. I want to speak from this place of simplicity, where I found my own truth, because useful acting techniques can only be about one's own truth, the truth one is experiencing in the moment.

The Michael Chekhov technique is a very free way to work as an actor. The material we encounter is immediately provocative and rewarding. I have been a practitioner of this approach for 30 years. I have been a teacher for nearly 20 years. I know most of the Chekhov teachers professing this work throughout the world today. It is not such a large community, and I am very happy to be a part of it. One striking thing about this group of teachers is that we are individual in how we see things and how we do things, yet the essence of the work shines in all of us. We stress the elements of the technique in different ways. We have all found what we believe to be the essential elements for our own work. We each go our own way in the technique. Chekhov's methods are actually limited in scope. His intention is to lead the actor to an inspired performance. The ways in which we can bring ourselves to inspiration are limited. This is actually a good thing because we can choose exactly how we will work and in what manner we will approach the role; and we will expect not to flounder.

I am now presenting an approach to this method, and I want to make it clear from the start that what you will find here is how the technique has come to me, how it speaks to me, and how I have made it my own. I quote Chekhov in these pages, but I choose the words that have caused a stirring in me. I do not believe there is one orthodox way of teaching or using this technique. I have seen some extraordinary teaching and it was new to me, things I had never seen nor considered before, but I recognized how 'Chekhov' these new things were. It is easy to know when it is Chekhov, and also easy to know when it is not. The purpose of the technique is to inspire, to find a creative state that is both pleasing to be in, and also full of the power of expressing oneself. Through the principles he offered to us, Chekhov expected each actor to find his own technique, his own way of working.

The result of the technique is also the thing itself. The technique is inspiring because of the very special demands it puts on the actor. I have chosen this as my way of working and I have looked into the entirety of it, but I have plunged headlong into the parts of it that especially speak to me and excite me. This is the material I will share with you. Some of it can be backed up by reading Chekhov's books, and some of it I developed out of principles I have found within the technique. There are also some other things I have encountered in my journey of physical theatre. I am not a purist, so I am always on the lookout for material which I recognize as being in the orbit of Chekhov. All of it has a truth for me and so I hope you will be able to find some truth in it as well.

Michael Chekhov was ahead of his time. He must have known that because he spoke about the *Theatre of the Future* and also about the 'actor of the future'. Fifty-four years after his death does indeed bring us into what he could call the future; perhaps he was speaking to our current generation. Artists nowadays have ears for an approach to working and seeing such as Chekhov's. With the proliferation of Eastern philosophies and practices in the West today, we are open to more concentrated approaches to working. We accept spiritual and energetic influences on life. The purely intellectual psychological method has already demonstrated its limits for performance. The conception of a human being as an energetic force is no longer an idea that needs defending. The mind–body connection is now commonplace. Modern humans have reclaimed this non-materialistic attitude towards themselves. This is the basic working stuff of the Chekhov technique. This book is intended to speak to modern actors of today about techniques that were devised for you.

Let's take a few steps back and begin at the simplest place, the physical presence of the actor; the actor standing before the public. What is true about this is that the actor is occupying space, either filling it with energy or not. If there is sufficient energy, then there is an interest on the other side; if not, then boredom or disinterest looms up. This is clear and can be easily demonstrated by these amusing yet profound words of Michael Chekhov quoted by Ms Deirdre Hurst du Prey, while she was teaching a class: 'The moment you are not alive on the stage, you are dead.' Now this is a statement that an actor can understand, because every actor knows the immense pleasure of feeling alive on the stage, and the profound pain of losing the audience due to a lack of energy.

Energy is a loaded word, it means many things to many different people. Let me define the word as I understand it, so we have a common ground to stand on. In my classes I define it as the force that moves the body, a substance that is neither muscle nor bone. It is the life force. It is called by many different names, a few of the more common usages are spirit, chi or prana. In order to work with Michael Chekhov's technique, it is essential that we have an understanding and an appreciation of this concept. Energy is the key that opens up the doors.

Imagination is another key because all our possibilities lie in this ability to imagine. A good starting place is to imagine that we are great actors. It's an image, an ideal picture of where we are going. It helps us to proceed toward our goal, it keeps us engaged. As actors we must be creative all the time. We have to believe that our job is to make art. Our art is a living thing and it is bound up with truth and reality and humanity, and the theatre. The theatre is full of hope, full of artistic illusions, full of imagination, and people. We can encourage the spectators to use their imaginations, they can become a creative presence there along with us. It is better when everyone is active and alive in the imaginary world, because the exchange between the actors and the audience is truly an exchange not just a one-sided assault from the stage.

It is my intention that actors, directors and teachers of acting will see this as a workbook. I hope it will be used as a practical guide to understanding the techniques Michael Chekhov has given to theatre artists. My appreciation of the technique as a dynamic whole came when I was able to discern the differences between the principles and the tools. This caused the components of the technique to fall into a clear and workable form, and allowed me to easily pick and choose the material I needed to engage in for a particular performance. In the book, *Lessons for the*

Professional Actor, where Chekhov is speaking directly to a group of actors in New York City, he states that 'every role requires its own technique', suggesting that certain elements of the performance are already understood by the actor so they need no attention. He also states in the *4th Guiding Principle*, that the technique is one whole thing made up of many parts, and that to engage one part is to engage all of it. This allows a very simple approach to creating a role.

Chekhov calls a great deal of the material we rely on in the technique as *intangible*. Because it is intangible, actors reading about it can easily become puzzled. Chekhov's book *To the Actor* is a great offering to actors but it remains a difficult book from which to work. He was unable to include the spiritual in his book about acting and still get it published. So something is missing. Through research into materials left by Ms du Prey, we can find many references Chekhov made to the 'spiritual' but these sources are hidden away in certain libraries, and only the serious researcher will take the trouble to locate them. The people who work with the Chekhov technique talk about, and exercise, and play with the thing that is missing from his book. It has come from person to person and it remains at the centre of the work. It is spiritual in the way that one feels one is engaged with 'something else', but not spiritual in a religious way. I want to address that missing part in this book so that actors can put it into play. I want to show you a way to the work that is reliable and clear. The more I have worked with this material, the simpler it has become.

Chekhov's techniques for acting are based in one primary point of reference. This point is movement. When we look at his system, we keep coming to this point. As we investigate the technique and return to this point, we find we are standing in a different place. Somehow the point is always moving. This technique begins with moving the body, because the moving body is what the audience sees. It is our front line of expression, it is our fall back. Becoming aware of the body as an instrument leads us to become sensible and sensitized to movement, even movements that are now super-sensible. As Chekhov said:

> Everything you do now consciously will become in time *super-conscious*, and that is our aim – to create super-consciously.
> (Michael Chekhov: *Lessons for Teachers*)

There are many ways to move, and many ways to perceive movement; it starts and it stops, it is both action and reaction, it forms and it destroys, it lifts things up and casts them down. It is essentially breath: the in and out of life.

Michael Chekhov's most conspicuous contribution to the actor is what he has named, the 'psychological gesture'. This is a very specific movement that is first imagined then executed as a means to excite the actor to play. It is in many ways a question posed by the actor, a question concerning the how of this or that expression. Receiving the answer is the initial difficulty of the technique and this brings us back to the body, which must become sensitized to the movements it can make.

Our instrument is the same body that carries on a life; it eats, and sleeps, and makes love, it laughs and cries, it dies. Experience is coming to us through our bodies as sensations. Our bodies record this as knowledge. We speak a language of experience we are comfortable with, using word pictures that are absolutely connected to movement. Perhaps we are too comfortable in that we have lost a connection to the original statements. What do we mean when we say, 'she fell into despair', or 'fell into confusion', or 'fell in love' or 'fell asleep'? How can these things be connected? Do we really fall into them?

Behold our common language of movement: We say we are either moved or not moved by things. We get behind them or we throw them out of our lives forever. No one likes to feel pushed into things, and we are quick to dump this on people, but sometimes it is sweet to be pulled along until we are joined willfully, at which point we begin to flow and we are picked up by this flow so our spirits are lifted until they fall again, when we may be induced to tear away or fade away. Or perhaps the fall has happened because we have been torn apart. Our hearts go out to others, or they break, our chins drop, we rise to the occasion, and swell with pride, we shrink in fear, or firmly stand our ground. We feel others out, put our heads together and touch upon the problem, sidestepping the real issue until we are able to draw conclusions and then finally rest assured, etc.

Movement is at the centre of these statements so it is essential we pay attention to movement, the movements we make, the ones made around us, and the ones happening within us.

Chekhov has said that there are two ways to work. There are two ways to concentrate and he makes it clear which way he prefers. Nothing will happen the way we want it to, unless we concentrate in the way he suggests. If we follow his lead here then something happens, we find we are immediately engaged with pure acting, and we recognize that we can go to new places within ourselves.

Very little here is new, After all, they are Chekhov's ideas. What is different is the structuring of the tools and the principles, and the inclusion of the dynamic force of energy. I have come to see the technique as conceived around archetypal energies. Concepts such as inner movements, radiation, atmosphere and incorporation of images are big tools. The only way we can successfully engage them is energetically.

The book is laid out in a very straightforward manner. The principles are separated from the tools to make it easy for the reader to reference them. The fifth section of the book, called 'Application', puts these elements together into a dynamic model. I have found a way to talk about the material that is both practical and simple. The Application section is made up of edited class transcripts. Certain parts reference the play, Desire Under the Elms, by Eugene O'Neill, while other parts are pure technique classes. My students and I are working together, and we have some dialogue about the exercises.

1

THE AIMS OF THE TECHNIQUE

Chekhov imagined the *Theatre of the Future*. He was convinced that it would happen and actors would come to meet it fully prepared. The aims of the technique speak about an ideal. Michael Chekhov is quite eloquent about all of this and these ideals are scattered throughout the books and teachings. A long time ago these ideals kept me reading his words and they have stayed with me; I do believe in them. They give us a picture that keeps the destination clear:

> The actor in the future must not only find another attitude towards his physical body and voice, but to his whole existence on the stage in the sense that the actor, as an artist, must more than anyone else enlarge his own being by the means of his profession. I mean the actor must enlarge himself in a very concrete way, even to having quite a different feeling in space. His kind of thinking must be different, his feelings must be of

a different kind, his feeling of body and voice, his attitude to the settings – all must be enlarged.

(Michael Chekhov: *Lessons for the Professional Actor*)

Our own 'I Am' is usually weak, but if we do the exercises of concentration, we will see that this feeling of 'I Am' becomes stronger, and we will feel as if we are centralized on our own spirit. The ability of concentration and the exercises, if they are done sufficiently and with the proper activity, will give this marvelous feeling of 'I Am'. With this 'I Am', we will begin to get our own being centralized, so our body will become centralized and our spirit will be centralized. This is the most beautiful thing, and especially for an actor who shows his whole being and nothing else on the stage. Then we will immediately become artists in the highest sense of this word.

(Michael Chekhov: *The Actor is the Theatre*)

Through concentration, the Chekhov technique leads actors to discover a power that is greater than the everyday sense of being humans. The real work of the actor is to transform personal experience into a universal and recognizable form of expression that has the ability to change something in the spectator. To simply reproduce a personal impression as it was experienced is not enough. As actors practising, we are saying, 'I am', again and again so that we can come to know the many 'I ams' that live in us. As an actor I find a way to say and believe in these words so that they can be a starting point for my work. The feeble 'I am' of everyday cannot be enough, I have to look for ways to increase this sense of self so that I can transform into other characters.

We live in an age where all of our responses to life are monitored, our thoughts and feelings are continually questioned and weighed into the scale of social acceptance. Generally speaking, we grow up in a world of doubts, apologies and yieldings. Pushing that all down, some of us decide to become actors.

Hopefully we have talent, or the natural ability to do it, because the talent of the actor needs attention. These techniques appeal to the talent. The instrument is the body with the voice, but we have to bring our talent into it. There is only so much we can take from work on ourselves, on our psychology, on our personality. We have to bring a clear and objective power to our talent so that we can interpret the lives written for us to act, or to act the lives we are creating in our rehearsals. Good intentions are never enough. Dance and speech lessons are not enough. A technique is necessary.

We trust in methods that use images to transform us, and believe that there is a radiant energy inside us and this energy can be formed and made active. If the joy of the actor is to give all at every moment, then there must be something to give, there has to be an inexhaustible supply of energy:

> When I try to imagine what the theatre can be and will be in the future, it will be a purely spiritual business (I speak neither in the mystical or religious sense at the moment) in which the spirit of the human being will be rediscovered by artists. The spirit will be concretely studied . . . [I]t will be a concrete tool, or means, which we will manage just as easily as any other means. The actor must know what it is, and how to take it and use it. We [will] know how to manage it, and understand how concrete and objective it can be for us. I believe in the spiritual theatre, in the sense of concrete investigation of the spirit of the human being. But the investigation must be done not by scientists, but [rather] by artists and actors.
>
> (Michael Chekhov: *Lessons for the Professional Actor*)

The actor looks to the essence of things. In the essence are found the building blocks from which he can recreate the world of the character. The details are created out of the essence.

Michael Chekhov was a very gifted artist; his technique was formed as a result of his ability to concentrate and to look at how he was concentrating and what he was putting his attention on. He saw what was at work for him. His early training with Stanislavsky enabled him to have a clear starting point, a new 'I am'. But his technique was his own way of working. 'I invented nothing he said, I have been observant, and discovered this is what I am doing when I act.'

The actor develops his skills in order to be capable of everything demanded of him. This development and the ability to create belong to what Michael Chekhov calls the 'creative individuality' of the actor. The 'creative individuality' allows the artist to use parts of himself that are not just the meaner more banal elements that make up his daily life, but rather parts of his subconscious, where dwell more universal and archetypal images.

In this way, the ego of the character is not subjected to the ego of the actor. The actor's creative individuality seeks an aesthetic union with the character, and will not allow the actor's personality to interfere with that process. With this the actor's work becomes an artistic creation:

> We have lost the whole poetry around our art, and it has become a dry business. The whole theatre has become so materialistic for us as actors: our attitude towards ourselves, our bodies [and] voices, our approach to the new play and so on ... Everything is condensed to the present moment, and even more to the events of the present moment, and even more to certain events ... The future theatre cannot go along this way of condensing and making everything dry. The theatre must go the opposite way, which is to enlarge everything: the point of view, the means of expression, themes for plays, and – first of all – the kind of acting.
>
> (Michael Chekhov: *Lessons for the Professional Actor*)

Artists desire to work from an inspired state. Yet inspiration is a fickle thing. The Chekhov technique addresses this desire. It aims to entice the inspiration to wake up for the artist. This is a bold claim that Chekhov makes again and again when discussing the method with his students. We begin with that promise, and by using the techniques, discover very quickly that we arrive somewhere within ourselves that is very new, yet very familiar. This creative place is fresh and available; it leads us to pure acting. Chekhov defines this pure acting as being able to happen without justification, without personal reasons, without psychology. Inspiration happens simply because we are actors, and we have engaged our actors' talent:

> We must never stop acting. We are always going on, and if we know it, our inner life, and power, and beauty as artists will grow, will show itself, and we will use our means of expression better and stronger than if we are under the impression that sometimes we are active as artists and sometimes not. If this seemingly simple and not very important idea is digested, you will see how much it will give you and disclose for you, and in yourself things may arise from within which you cannot get in any other way than to change your point of view and get new conceptions of yourself and your art.
>
> (Michael Chekhov: *Lessons for the Professional Actor*)

The chart on the facing page describes the processes and progressions involved in acting with the Michael Chekhov technique.

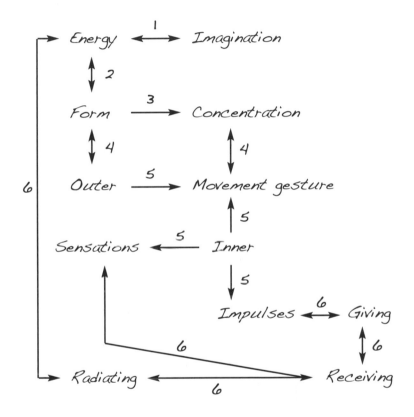

2

THE PRINCIPLES

Chekhov has given us what he calls the 'five guiding principles'. These guiding principles should lead us through the process of acquiring and developing the technique. He tells us we have to train in order to have a general technique, and then we apply specific techniques to a role. The first five principles listed below will be the five guiding principles.

In all there will appear more than five. My approach to the technique is to take the principles and apply them by using the tools. In my quest to master it, I have differentiated certain elements of the technique as *dynamic principles*. These principles are the reliable power for us. They are points of reference we keep returning to. They can sustain us as we sustain them. The principles are like prisms, the tools, like light passing through them. All the colours we experience, we use to express ourselves.

1 THE TECHNIQUE (ACTING) IS PSYCHO-PHYSICAL

The body and the psychology are one thing. The body is developed and trained so that it becomes sensitive to this connection. Movement is not gymnastic but psychological in that it affords us the experience of states and conditions of being. The good result of movement exercise is a fit body, for us it is a good benefit but not the aim. The body must act as a sponge to absorb psychological values or qualities from the movements. These movements are repeatable and can be used during the rehearsal to anchor key moments of the scene within the body. The psychophysical exercises devised by Chekhov aim to develop the two powers of concentration and imagination in tandem. Conscious movement involves much more than muscles and bones. With proper concentration we experience movements so as to re-educate ourselves. We become familiar with the actual movements that surge through us.

Our normal lives prevent us from recognizing these movements, due to our habit of passing experiences through the intellect. Conscious movement helps us become aware of clear impulses that lead us through our daily lives. We see them for what they are, where they move, and how they move. Then we learn how to access them and follow them when they naturally occur within. As the concentration develops we can then begin to imagine these inner movements as happening. We make them happen. We develop the power to change things for ourselves in our lives and in our art through the imagination of movement. Action and reaction, giving and taking, laughing and crying, living and dying, can all be seen as movement. And as they move through us they will move us to act.

If the impulses that spring up and die within us are not followed or actively resisted we will naturally lose consciousness of them. Our focus as student actors here is to reacquaint and then to reanimate that which we have let atrophy within. With the body flexible and soft, absorbent and expressive, we find again the pleasure of movement, and the surety of a physical wisdom our thinking could never achieve.

2 INTANGIBLE MEANS OF EXPRESSION

The promise of the technique lies in the idea that the most effective and powerful means at our disposal are intangible. They are present only when the concentration is active. These intangible means include *atmosphere, space, radiation, relationship, inner movement, imaginary body, imaginary centre*. When the concentration fades, so do these means; they cannot exist without an imaginative and concentrated effort. They are intangible because one cannot put a finger on what it is. We know when it is present, and we know when it is not. When it is present we receive what comes to us and use it as our means of expression.

3 THE CREATIVE SPIRIT AND THE HIGHER INTELLECT

There is a spiritual element to this work that must be acknowledged. This spiritual element is not religious. The creative spirit (imagination) is differentiated from the reasoning mind. The creative spirit as Chekhov talks about it functions within the artist by making *one thing* out of a multitude of things. This faculty can grasp understanding through archetypes and through a desire to find wholeness. It is the quick creative function of synthesis. The creative spirit is capable of working in this way; the rational mind works through analysis. Analysis separates and divides, whereas synthesis unifies and brings together the many disparate parts we encounter in the preparation of our composition (role). The work is intuitive, the results come and are actively invited into the consciousness, so they can be experienced and expressed.

4 THE TECHNIQUE IS ONE THING: IT AWAKENS A CREATIVE STATE

There are a number of components in the technique. Each one of them needs to be examined individually and practised thoroughly. The actor learns to distinguish between the components. The creative spirit connects them to each other. Each component opens a door to inspiration. Through concentration we can activate one of them. Because we have a familiarity with all of them, this one component will cause any or all of them to become active.

5 ARTISTIC FREEDOM

The technique promises artistic freedom. Chekhov suggests that this principle be engaged through a dialogue with the technique itself. It is the way to know how to work. Rehearsing and performing is the work of the actor. But how will he do it? If he has a method to use, and it is made up of different parts, then it is necessary to confront this method and to inquire of it which part of it speaks the loudest to him. Which part of this technique gives him the freedom he seeks as a performing artist?

3

THE DYNAMIC PRINCIPLES

1 ENERGY

Nothing is effective without energy. Real energy is inexhaustible. Energy breeds more energy. Everything radiant is energized. Energy is full of life. Energy supports life. Energy moves life. Raw Energy is formless. Giving form to energy makes it creative. The human body is a conduit for energy. The many parts of the physical body have corresponding energetic parts. The shape of the energy within us is the same shape as the body.

2 IMAGINATION

William Blake, imagination's fearless champion, wrote, 'What is now proved was only once imagined.' He saw the imagination as a divine and active gift in human beings. It is our connection to pure energy. When we begin to work as artists, it is the first thing we appeal to. If we persist in our endeavours, we find a way to connect our technique to this most precious activity. It is central to Michael Chekhov's approach to acting. It is always addressed and daily developed. It becomes the stepping off point for students of this method. As student artists we are taught to imagine and then to cross the threshold of our daily lives into the world of the creative artist. Quite literally we learn to step into another world where our inner powers of concentration and imagination wake up. Upon crossing this threshold, we continue walking in what Chekhov called the 'actors' march'. This exercise is an affirmation. The words empower us to want to use the instrument we have to our fullest capacities:

> I am a creative artist.
> I have the ability to radiate.
> Lifting my arms above me I soar over the earth.
> Lowering my arms I continue to soar.
> In the air moving around my head and shoulders I experience the power of thought.
> In the air moving around my arms and chest I experience the power of feelings.
> In the air moving around my legs and feet I experience the power of will.
> I am that I am.
>
> (Michael Chekhov: *The Actor is the Theatre*,
> Deirdre Hurst du Prey)

This affirmation of ourselves leads us easily into how Chekhov envisioned the ideal actor. It gives us pictures of the actor moving, and then moving without moving. Engaging human functions and realizing that everything we need as actors is within us already, waiting to wake up. We also see it is all around us for the taking, if we are open to receive it. We are asked to envision this ideal as already achieved, so that our working towards it comes with pleasure and foresight of success. It is an embodiment of Blake's words 'What is now proved was only once imagined.'

The use of imagination and the image gives the actor a freedom to reach beyond his personality. It allows him to be led by a power that is continually expansive. It is brought into play by concentration.

3 CONCENTRATION

Concentration is the key activity in realizing anything of value. To concentrate does not mean to think harder about some thing. When we concentrate we *send* ourselves towards an object or image. Once we become one with the image we can feel its quality, sense its personality, receive impressions and impulses. A concentrated artist makes a huge impression on the audience. Art is not really possible without concentration.

When we are looking at something we are attracted to, we somehow feel that we are moving towards it; it pulls us in with real attraction. This is a pleasing sensation. It is a concentration that is happening on its own. A willed form of concentration is possible when we can send ourselves to whatever we choose, in order to become one with it, and know, in an inner sense, what this image or object is. In this way we can become psychologically identified with the image.

4 INCORPORATION

Acting is completely wrapped up in the body. In order for us to experience the images, and express the things we are seeking, the images must be incorporated. They have to be put into or onto the body. Once the image is successfully incorporated, we play the instrument to express what needs to be expressed. Incorporation is the direct consequence of the concentration.

5 RADIATION

The inner work of acting, the knowledge, the feelings, the actions must in the end come out to touch the audience. Whatever is living within us can be sent out in an energetic wave. Radiating is an activity that will accompany an actor who is inspired. It can also come as a result of willing it. It produces pleasure for the actor to do and for the audience to witness. It simply is a sending out beyond the body what is alive within the body. It touches the audience because it actually goes out to them.

6 EXPANSION/CONTRACTION

Two dynamic forces that impact the natural world are funda-mental principles of human interaction. It is very easy to see the effects of expansion and contraction in the world that surrounds us. When we observe the differences between winter and summer as two polarities, we will see that to move from the depth of winter to the height of summer is a slow and steady expansion, while the reverse is true, a steady contraction from summer to winter. Everything responds thus, and all creatures understand the dynamics of growing and receding. The body and its parts move in the same way, muscles, lungs, eyes, ears, blood vessels, expand and contract. The emotional life too is in a flux controlled by these forces. The world of thought, as well, depends on being open to certain things, and closed to others. These kinds of things sound obvious and simple when we talk about them; we are such busy creatures occupied with the many complex details of life, that we tend to forget the most basic things.

There is no way that another person can convince me of his point of view if I am closed to it. There is no stronger indicator of resistance for that person than me in the act of closing. This contraction of mine signals the other person to do something, to move either outwardly or inwardly.

A physical acting technique cannot really be taught without working on this as an artistic principle. It must be investigated by the student actor, explored, tested and applied as much as possible. It is so useful and practical.

7 SPACE IS DYNAMIC

Space filled with possibility is a powerful ally to the actor. Chekhov tells us the surrounding space is just asking to be engaged. If we imagine it to be thick then movement slows down. If we make it warm then things appear closer because bodies are attracted to this kind of heat. Filling the space with coolness creates distance and also a kind of clarity, because this is how things are for the body in a cool space. Make it fragrant and we begin to open up; acrid and we close down the senses. It is not about pretending that the space is cool or warm, nor do we want to show that it is cool or warm. The *imagination* supplies the *atmosphere*, then we receive the cool or warm onto our bodies and let the reaction guide things for us. The body is consistent and reliable, also predictable and expressive.

A company of actors working together to imagine the space filled with an *atmosphere* can make an astonishing impression.

8 DIRECTION IS A FORCE

Movement happens in space, usually in one direction at a time. There are six directions we feel as real force. The dynamic directions are: Expanding and contracting, forward and backward, up and down. We immediately recognize what they are, and what each direction means to us. We quickly understand how we can use it. These directions are forever corresponding with human interactions. They are connected to all the *sensations*, and to all the actions actors involve themselves with. Awareness of direction is an easy and valuable lesson to learn. It forms and informs the performance.

9 POLARITY

Events happening between two poles cause a vibration to occur. Contrast allows things to stand out from one another and in relation to one another. Any work without contrast is dull. The edges are not clear and everything seems to have the same value. Polarity makes things instantly interesting. In the Chekhov technique we are always looking for ways to use it. While exercising, it is frequently explored, and always considered while rehearsing and performing. The beginning and the end must appear as a polarity. The more polarities we find, the more exciting our work will be.

10 QUALITY

How to do it, is an artist's question. How do we interpret the role or the play? How do we make form and sense out of the given circumstance? How do we order the talent to express itself? Quality can turn one thing into a multitude of different things. It can transform a kick into a kiss, or a seduction into a murder. How it is done is the joy of creating, the pleasure of being an artist. The quality is how something is done. The quality speaks directly to the feelings of the actor. We often use words like tender, soft, strong, courageous, sluggish, vibrant, proud, quick, heavy, etc. to describe characters. This is how they appear to us and these words help to identify the essential aspects of them. We can also make our movements, both outer and inner, using qualities and they will say something very unique depending on how they are done. What is exciting about the technique is that we can go outside of ourselves to find source material for our creations. We learn to appreciate the quality of the world around us in a very practical and creative way. We can look at the quality of objects, images and people. We can perceive something truthful about them, and directly experience a feeling for them by penetrating into an appreciation of their quality. We express ourselves by engaging this question of how. Through quality we find the ways to do the things we have to do.

11 THINKING, FEELING, WILLING

When we look for the simplest things about being human then we find a way to work as actors. It is out of the simplest functions that deeper complexities can appear. Simple ideas are the best ones for us to look at, because they are possible to understand, to feel, and to do.

We as creatures are capable of three clear functions and it is the interplay of these three things that make up a life. As humans we are very proud of the fact that we can think, it is our domain, and it can be viewed as a function. We also have the ability to feel, and our feelings often rise up in us as a result of our thoughts. The next function is action, doing, expressions of will power. These can also lead to feelings, which in turn can excite more thoughts, which could then cause more actions or feelings and then again more actions and then some thought, and so forth. This view is limiting in its scope, but at the same time includes all possibilities. These three functions become wonderful containers to put the material of the play into: the characters in relationship to each other, the words they say, the way in which they say them or hear them.

These ideas are not new nor can we ascribe them to Michael Chekhov, but Chekhov did offer this picture of the human being as a principle to work from. In fact, when discussing the play itself, he suggested we look at it as a living human being and try to discover what are the ideas or thoughts of the play, what are its atmospheres which could also be called the feelings and what is the will, the actual things that are done and seen by the audience.

The very first thing we need to know is: What is the difference between a thought, a feeling, and a will impulse? Thoughts are real things that occur quite locally within the body. We do not think with our legs because they already know what legs are supposed to know. Thinking is a process of working through something that we do not already know. If we knew it then we would not have to think about it, it would simply appear as an image in the mind, like a chair. This isn't an activity, it just is. If the chair is unsteady then we might be compelled to think about why this is so, or how we could stop it from being so. Now we have to think about it. Or perhaps we might try to remember something, or calculate something, or invent something. These are different thoughts which take place in the head.

The chest is where the heart lives, and the world of feelings has always been linked with the heart. In all languages, hearts are broken and mended by love. We do not as a rule think with our hearts, but we can speak from them and listen with them. Movements originating in the chest allow the actor to connect with the feelings, or the feeling life of the character; they also allow the audience its own sympathetic resonance to the moment.

The world of appetites and sexuality located lower in the body can be the starting point for basic expressions of the will. I want, I take, I give, I reject: these things originate low and sure within the body with the pelvis and the legs. Attention paid to this location of the body excites bold clear actions or doings.

All this simple business may appear primitive on the surface. But Michael Chekhov's technique is never on the surface, it is always deep inside the core of the actor. And so we connect to the impulses moving through the centres of thinking, feeling and willing, we say yes to what is universal and human in all of us, and we make conscious use of it.

12 THE FOUR BROTHERS

Every great work of art possesses four common traits, necessary elements towards a satisfying perception of the art. Each one complements and informs the others. These are quite tangible things for the actor to look at, and to work on. And they are the following:

13 A FEELING OF EASE

It is possible to see a feeling of ease because it is apparent when someone executes a task or an action with ease. It is also apparent when there is a lack of ease. No feeling of ease is easily perceived and it has a detrimental effect on the audience, especially if what they are looking at has an element of risk or danger. The audience does not want to really worry about the actors on the stage, at least not in the same way that they worry about the characters. The safety of the bodies on the stage is not part of the bargain that was entered into. But the actor finds this ease as an inner thing much like a feeling. It begins from the desire to have it and it comes because we will it, and we know that it is one way to make our work artistic.

14 A FEELING OF FORM

Everything that needs understanding uses form. A feeling of form can really be experienced as a feeling, because our human body is a form. We begin our understanding of this through the physical body, how it *feels* to be in the body. Being in the body is something few people actually experience beyond the sensations of injury, illness and distress. We learn to feel this human form, and know the particulars of it as well as the unity of it. This form moves forms and these movements have a beginning, a middle and an end. Here are some other forms for our consideration: the play, the scene, the monologue, the stage, the scenery, the props, the sound, our fellow actors, etc.

15 A FEELING OF BEAUTY

This feeling of beauty is not an easy one to feel, it has so many values attached to it. Perhaps it could be looked at as meaning *authentic*. Chekhov said that animals in nature are beautiful simply because they are always true to themselves. A butterfly on a flower, a tiger stalking a rat, these things hold a fascination for the viewer. The feeling of beauty is somewhat elusive. We cannot find it by trying to be beautiful because this only produces the opposite effect, but we find it in a most natural way by playing all around it, and also with the other brothers of Ease and Form.

We are capable of making ugly things but this has to be conscious work. If we can approach our work with a feeling of beauty, then we will not make unconscious ugliness. It isn't showing off, it is exactly the opposite; it is being appropriate, virtuous and positive.

16 A FEELING OF THE WHOLE

As actors we are never alone, we are part of something which is greater than we are. But we are one thing, and the whole of what we do is one thing, and the part of what we do is also one thing. Everything works as one, and each moment is one thing and it points towards the unity of the entire composition. Even in time, what will be seen in the end needs be present in the beginning, and the beginning be present in the end. Each piece taken out reflects the whole.

4

THE TOOLS

There is always a certain 'what.' [For example,] the play is 'what,' [and] we have to deal with our parts as 'what.' There are two ways in this 'what:' one is leading to 'why,' and that is pure science. When we take a play and try to discover 'why' the author has done this or that, we will never be able to act it. The other way is 'how,' and that is our way as actors.

For instance, if we know how to become jealous on the stage without knowing why, then we are artists. The more the materialistically minded world forces us to go the way of 'why,' the less we are able to develop our abilities and talents. This 'why' is very widespread in art in our present life.

If you ask how can I know 'how' if I don't know 'why,' I would say that it is a very materialistic question, because 'how' is the mystery of art. It is the secret of the artist who always knows 'how' without any explanation, any proof, any analysis or psychological abilities.

(Michael Chekhov: *Lessons for the Professional Actor*)

1 TRANSLATING THE INNER EVENT TO AN OUTER EXPRESSION

For an actor to have an effect on the audience, this actor must be alive; a dead actor has absolutely no effect. What makes an actor alive? The first part of being alive is to really be alive: to have life within. If one were to compare a living body to a dead body, one would instantly become aware of one thing; the living body is animated. Implicit in movement is a kind of life. Within the living body there appears to be a vital force, which is clearly absent from the dead body. We cannot really see this force but we can see its effect. We could call this vital force an energy, a life energy which maintains the life in the living body. It allows all movements, voluntary and involuntary, to occur. The quality of the energy is a guide to its health. It is possible to look at a living body and see the quality of the energy within, and then to say that this organism is strong or weak. Without the energy's influence the physical body has no support, it falls in fact, and immediately begins to decay. This energy exists, but until I can recognize the energy within, it has no form, only force.

The second part about being alive as an actor is to be able to appear fresh and spontaneous, completely involved in the circumstances of the character as if for the first time, responding with fullness and truth. This, of course, is the aim of every acting technique. Michael Chekhov's technique promises to create the second part by working with the first part.

With a small amount of imagination applied, we can begin to give this vital energy a form. Let us suppose that the form it takes is a kind of inner replica of the physical body. The human being has a body; it is a form. Inside this form one can imagine is another finer body (form) composed of life energy. The physical body moves, this is the actor's means of expression; she moves in response to the world, she moves towards things, or away from them, with them or against them. Sympathy and antipathy are the causes for her movements, also desire or will.

It is possible to imagine a movement and have an experience of the movement simultaneously. It is simple to do, and as soon as we have an experience of this then so much can be understood. If I tell myself to move my arm up and down, and I do it, this requires nothing more than desire. It is an ordinary movement, something I do everyday. I can continue to practise this movement so that I become very familiar with it, then can stop the movement all together and begin to imagine that I am moving my arm up and down. This particular imagination is not visualization, it is a *movement imagination*, as I intend to have the experience of this particular movement without actually moving my muscles. Energetically this *inner movement* is as much an event as the actual movement. But it belongs to me, the actor, because it is invisible. The appropriate inner movements are a means to excite within me the expressions required of good acting.

We can begin to work with ordinary movements in order to exercise our ability to experience movement without moving the visible body. The purpose of this is to feel as if we are moving. Once this is incorporated, it becomes quite a pleasurable and free place to work. It is limited only by the images the actor is able to find and put to use.

This way of working is best suited to talented actors because the principles involved are a direct appeal to, and an enriching of the talent within the actor. By appealing to the talent, and not the psychological history of the actor, the technique opens convincing possibilities to choose from that are no longer personal. It enables the actor to have a real-time inner experience right now, not a reconverted memory. It is called an *inner event*. What the audience perceives, the *outer expression*, is the response to this inner event. They do not know the actual cause of the response, and they believe it to be the circumstances surrounding the character. The talent of the actor allows a connection between two different occurrences (inner and outer) to take place, because the talent of the actor is in a state of giving to the audience.

It is incumbent upon the actor to know the outer circumstances surrounding the character very well. During the rehearsal process, different inner movements are being experimented with and laid down as sign posts or anchors for the performance. In performance the concentration is heightened, the actor really appears to be living freshly, spontaneously, completely involved in the circumstances of the character as if for the first time, responding with fullness and truth. Yet it is the image that is moving the actor night after night. The inner event, generated by the image, causes certain impulses to pass through the body of the actor. Following and/or resisting these impulses creates the behaviour of the character, the outer expression of the actor.

This basic operating principal, the inner event is translated to the outer expression, is the key to understanding how to use the techniques of Michael Chekhov. The training is always pointing to this ability of making an identification with an image, and it continually confirms that movement is essential to living things.

2 SPY BACK

When we are properly concentrated, then we are one with our images and our intentions. The necessary things are moving through us. We are being creative. We need to be present enough to get out of our own way and enter the flow that is given to us. This is not the moment to analyse it. When we have finished, and this is particularly true for exercises, then we can look back at what happened, we can evaluate what was working and how it was working in us.

We want to be led into our work by our imaginations and not our intellect. We are normally led into everything by the intellect. It is used to being in charge of our lives. Because it is in a position of authority, the intellect does not relish letting go of the reins, so to speak. Chekhov said the intellect is a kind of enemy to the artist; he called it the 'little intellect'. We know this little intellect, it is the critical, judging, discerning and divisive part of us. It protects and guides us in many things, but does not help us in the creative state. So we try to subdue its influence by engaging the imagination.

When we have finished an exercise, we can look back at what we have done, and benefit from pressing the little intellect into service. Chekhov called it spying back. If we can engage the little intellect here, then it becomes satisfied that it is involved with our process. We lead it away from the real work of creativity, which is intuitive, impulsive and physical. The spy back is a good rehearsal device, and a brilliant learning tool. In class these spy backs are shared and the students grow together in understanding the value of the exercises. Questions to ask in the spy back:

- What was I concentrated on?
- What does this movement mean to me?
- What is my experience of this?
- Where is my connection to this?
- Do I recognize this?
- Can I do it again?
- Where can I use it? etc.

3 ENERGY: THE LIFE-BODY

The life-body is an expression I encountered some years ago. Michael Chekhov did not use this term in his writing, or in his teaching. I have adopted it as a way to describe the inner energy that we play with while practising the technique. As a teacher I have come to use this term, because I believe it describes succinctly and perfectly what this elusive and intangible force is. It is necessary to call it something, because I refer to it all the time. We could also call it the energetic body, but I prefer to call it the life-body because it gives us a certain kind of picture. This picture allows us a way to give form to the energy and it also provides the life that we seek as actors. We need to feel that we have bodies and that we can use them. The same is true of the life-body.

Stand still and feel your feet making contact with the floor. Here is an experience we are so familiar with that we hardly notice any more we have feet. Not until I step wrongly off the curb and twist my ankle do I ever tell myself that I have an ankle. But after the curbstone twist happens, I say to myself with every step I take, 'I have an ankle'.

Allow standing on the floor to become the experience that it is right now. When you can feel your feet making contact with the floor, then you can rightly say, 'I am present in this moment'. It is so because your attention is with your standing and that is what you are doing. Normally we are not actually present, but drifting with our thoughts.

Exercises should begin with feeling your feet making contact with the earth. This is a signal to yourself that you are present and ready. Auditioning, rehearsing and performing should all start from feeling the feet.

Exercise 1: Finding the life-body

Lift your right arm so that it is pointing towards the sky above you. Then let it return to just hanging naturally by your side. Repeat this a few times using each arm; see if you can feel what it takes to make this movement. Now imagine that inside your right arm is another finer arm made of energy, it is there within the physical arm, it has the same shape. Now try to lift the energy arm first, before you lift the physical arm. The physical arm will just naturally follow the energy arm up so that you will be pointing to the sky above you. Now begin to lower the energy arm and let the physical arm follow it down.

This can be done in two ways. You can imagine you are doing it by visualizing yourself doing it, seeing an imaginary arm rising followed by the real arm. You can also imagine you are *doing it by feeling* that you are lifting an inner arm followed by the real arm. There truly is something to feel here, and nothing to see. Put your attention on *feeling* it. It is much more work to try and visualize it and little reward comes of it. All the exercises need to be experienced. These first few exercises are focused on waking up the energetic connection so that all the other exercises can be approached with the attention on the life-body.

Exercise 2: Timeline

Easily lift your left arm as if to point at something on the horizon. Feel all the way to your fingertips. Then imagine that you can reach past your fingertips into a space just beyond your hand. This reaching is not a physical stretching. It is an energetic reaching. You must *believe* that it is possible to do this with the imagination, then it becomes a fact. The body, and especially the left arm, should be easy and light. If you feel that you are tense, then release the tension. Now you are reaching a bit beyond the body, you are *radiating* into a space that we could call 'tomorrow'. Continue and reach beyond that into 'next week', then reach even further through the opposite wall into 'next year'. *Feel* that energetic rays are leaving you in a straight line and radiating out beyond the wall of the room you are currently in. As you continue to look forward, lift your right arm into the space behind you. Feel that you are reaching beyond your fingertips into 'yesterday'. Please do not get distracted by what happened yesterday, it is not the point of the exercise. The point is to just *send rays energetically* into 'last week' and 'last year'. Focus on the line of energy you have just created, running from last year through you in the present time, and on into next year. When you make a contact with this line, say 'yes' to it. Now lower your physical arms, but leave the life-body arms where they are, they are a part of the line that help you to feel it. Square your shoulders. Stand there comfortably feeling the line, then begin to walk forward on the line as if to walk into the future. Stop. Walk backwards on the same line towards the past. Stop. Find the present time. Lift the physical arms up to meet the life-body arms. Consciously take back the line from the past and then take back the line from the future, lower your arms. End. Spy back. Repeat.

This exercise is a very clear way of understanding that we can move energetically, that we can move beyond the physical confines of the body, that we can *radiate* in one direction and then in two directions. The line itself is very pleasurable. It is quite possible to take this out of the studio and to walk for a distance on the line. Walking itself takes on a new feel, a real *feeling of ease* certainly appears. Try it when you are late and must walk quickly somewhere. You will be most surprised how things change in your journey. Once you have done it a few times, your body 'knows it' and you no longer need the big set-up and can just spontaneously project a line in two directions, then you can walk on it.

There are other exercises to help us feel the life-body.

Exercise 3: New eyes

This exercise is best done in a large clutter-free space. Stand up and imagine that you have eyes on your shoulder blades, and as you stand there, look behind you using your new eyes. Your normal eyes require no attention from you. The world *comes to you* through your normal eyes, you do not have to go to it. Focus your attention on looking backwards or behind you with your new eyes. In the front the world comes to you, in the back you must 'look' for it. Begin to walk backwards without turning your head or your normal eyes from side to side. If you are working in a group, really try to feel (see) when you are about to bump into another person. By looking back with your new eyes, you will be able to negotiate your way around each other. If you are alone, then stop just before reaching a wall because you can feel (see) it with your new eyes. Then stop using your new eyes and walk backwards. Is there a difference? Standing still, look back with your new eyes again. What has changed? Continue to look backwards with your new eyes but walk forwards. Stay awake and aware of what is happening in front of you, but continue to put your attention on your new eyes. Be present in the room now. If you feel that it is becoming trance-like, then stop and gather yourself before going on again. You need to develop the concentration so that you can do the exercise and not be lost by it. Always make yourself present. It will not serve us to float away in a trance. We desire to work consciously. Make eye contact with your normal eyes to the people you pass, this will help you tremendously to stay present. But stay involved with your new eyes.

This exercise delivers a genuine and unique sense of being a three-dimensional figure in space. It gives us a direct sense of having a back and that this back is in a relationship to the *backspace*, which is different from the front space. The backspace is a concept my teacher stressed. If we are connected to the backspace, we are in a place of power. When we 'light' up the backspace, we instantly feel more important. We feel ourselves to be occupying more space, and the things we do will have more consequence. Weakness and power as choices become understood in a spatial and physical way. The New Eyes exercise helps to develop the concentration. It is also helpful in the accumulation/repetition process, while developing the psychological gesture.

It is an exercise I introduce in the first meeting, because it allows us to experience ourselves expanding, and gives us a new and energetic starting place.

With new eyes you perceive a different sense of having a body, and that the body is in a space.

Exercise 4: Feeling the life-body

Walk around as if you are a character in a cartoon trying to sneak past a sleeping dog. Walk in a large and exaggerated way so your concentration is completely wrapped up in moving silently and largely around the room. Try different tempos and make your steps varying lengths. It is a very child-like thing to do. Play, and enjoy what you are doing. Sneak. After a few minutes, stop, then walk normally, but as you are walking, imagine that you are still walking like the cartoon. You can also just stand still and imagine you are walking like the cartoon. As you do this, really try to imagine the movement. It is very important to locate the imagination in your muscles, or your life-body muscles. Then you will feel like you are actually doing it. Stop. Spy back. Did anything wake up in your body, any sensations, or impulses, images?

Exercise 5: Tricking yourself

Walk quickly forward and tell yourself that you will turn right, keep reinforcing the idea, then physically prepare to turn right, but at the last possible moment, turn left. Try this many times in both directions.

While walking quickly tell yourself to stop, prepare to stop, but do not stop.

Or being stopped, fully prepare to go, but then do not move. Really try to trick yourself. Spy back.

This produces a strange sensation where you feel yourself separated in two. The energetic body is going right and the physical body is going left. The sensation is very brief but it becomes clear where intentions are located, and how easy it is to follow what is already in motion. The head rules in this exercise. Now we can feel how awkward it is when we do not follow the energetic body.

4 STACCATO/LEGATO

Mostly everything we do in our normal lives (except sport or physical exercise) is to deny that we have a body. If we always move in a normal tempo, it becomes difficult to get any physical sense of what we are doing. Repeated movements in a normal everyday tempo have a tendency to numb our experiences. If we speed up our movements, we *sense* that we are moving because it is a bit more difficult to do them. When we slow down our movements, we become very conscious of moving because it is taking more time to accomplish things. The point is to become conscious of movement. We have to have 'knowledge' of it. The actor's biggest problem is that her instrument is also the same thing that carries her through life. Everyday life makes us shut out some consciousness of moving. The initial intent of this work is to become re-acquainted with our natural physical reactions, which we *feel* as impulses and sensations. The next part is to discover how certain movements can animate the same natural impulses and sensations.

Tempo is a clear gauge of a character's core. There are two types of tempo to work with, inner and outer. Inner tempo is the speed in which the inner life moves. A slow thinker, or a quick hot feeler, or a sluggish but determined will: all that begins inside the character. Outer tempo belongs to what is called business, the physical doing of things. We can practise many of the movement exercises and vary the consequences by changing the tempo. Staccato is quick movement with sudden stops and starts. Legato is slow (not slow motion) and has no clear stops. It might stop, but then again it might not stop.

Exercise 6: Staccato/legato

My teacher, Blair Cutting, a student of Michael Chekhov, followed Chekhov's class plan. He began each class with the exercise of staccato/legato. It was of singular importance to him. I have been doing this exercise for 26 years and it is still as fulfilling as it was when I was a student. I have come to believe that the whole Chekhov technique is in this one exercise. It is an exercise that can be done in many ways, and with different focuses. The basic exercise is as follows:

Stand in present time. Know that you will move in the six directions of right, left, up, down, forwards and backwards. You will move in one direction at a time. You will make only one movement and this movement will be repeated a total of 36 times. Begin the movement by turning to the right and lunging onto your right foot, stepping on it taking all you weight there. It does not need to travel far, a short lunge is enough, just a real commitment to the direction of right. So, you are completely facing in this direction from your toes to your face. While you are making this lunge with the lower half of your body, imagine that you are holding two tennis balls, one in each hand, and you will throw these imaginary balls underhand as far as you can. The final position will be all your weight on your right foot facing completely in that direction, your arms fully extended in front of you, the palms of your hands facing downwards. It is all done in one efficient movement, one large gesture of throwing while lunging to the right. So now that you have moved your physical body in this gesture, what remains is for you to send out your inner energy in the direction of right. The energy should radiate out of your fingertips, your face, your chest and your knees. Try to throw it through the right-hand wall you are now facing. The physical movement should be done in the tempo of staccato (quick with stops). The radiation continues briefly then you return in staccato to the starting position. It is important to return to this position and be present in it, as if you never left it. So you have a complete commitment to moving towards the right, a throwing gesture, which helps you to throw out or radiate your energy, then a return to a clean starting position as if you had never left. Now repeat the same movement, only do it to the left, onto the left foot, committed to the direction of left, radiating in that direction, return to the clean starting position. From this position you will now throw upwards towards the sky, commit to the up direction, lift your face in this direction, and radiate, return to the clean starting position. Then throw everything down into the earth, bending your knees, radiating downwards through the floor, head facing down, return to the start. Now lunge forward onto your right foot and throw in that direction, radiate, return. Now step back onto the left foot, throw underhand towards the back, radiate, return. This is one cycle completed, six directions all done in the tempo of staccato. Repeat the cycle in staccato one more time. Then repeat the same cycle in legato (slow with no stops) two full times. Then repeat one time in staccato, and finally one more time in legato. The whole exercise should take no more than two minutes to complete.

Mr Cutting suggested doing this exercise on the stage in preparation for a performance, before the audience is admitted. I did that then, and still do. This is a wonderful way to warm up the instrument. It also allows you to fill the space with your energetic self. It is a creative act to do with other actors, as it helps the ensemble feeling. I cannot begin a performance without it now. It is a kind of cleansing as I can throw off unwanted stale or negative energy that can insidiously interfere with my best intentions as a performer.

The exercise also helps us to ground these two tempos within the body, thereby incorporating a dynamic understanding of character and quality. We can discover a great deal with these two tempos if we see them as viable means of expression. The polarity between them gives them significant consequence.

5 THE SENSES: EXPANDING AND CONTRACTING

Expansion/contraction is a principle but it is also a tool because it is something that can be done. It is an inner event and if we follow it then we are involved with the principle. Expanding or contracting is very specific activity that we can localize anywhere we choose. It is possible to expand the body, or the organs, or the senses, or the space. It always delivers great rewards when we are engaged with it. Experiencing the life-dody growing or shrinking fills us with the ebb and flow of vitality. Are we growing in the face of something or somebody? Are we shrinking? These are very simple questions that are easily answered, easily known, and the inner activity can be easily done.

Exercise 7: Relationship to people and objects

Make a soft fist with one of your hands. Look at the fist and slowly open your hand, as you do this tell yourself that you are growing. Try to experience it as a growing, a filling with vitality, power, effectiveness and strength. When you have reached the end of the movement, when your hand is at its most open, then begin to contract it into the soft fist again. Tell yourself that you are shrinking or withering. See it as shrinking, and try to feel that because of this movement a certain vitality is withdrawing from you. Play with these two movements for a little while, watching them and feeling them. Use both hands, and then add your arms. Do this easily and softly and you will soon feel what expanding is, and what contracting is. When you have this understanding, then you can begin to play with it in your imagination.

Tell yourself that you have the sense of taste. We know that we have this, but we take it for granted. Saying it is almost like discovering it, we come to appreciate it because it is new. We know that our sense of taste is located on the tongue and we can taste because we have taste buds there. Imagine that it is possible to expand your sense of taste. This activity, as you will find out, is a familiar one. Sustain the activity of expanding now and follow the inner activity. Whatever impulses come to you, allow yourself to follow them. Play as long as you are able to receive something from the expanding. Approach some object and see what kind of a relationship you have to the object. Now try the opposite and contract your sense of taste. None of this is actually about tasting, so don't get confused here about the exercise. It is literally about contracting your *ability to taste*. If we make this about tasting chocolate or anchovies, then there is nothing we can sustain because we put our energy into recalling a specific taste and maybe we will get lucky and actually recall it. But we can sustain a growing or a shrinking and it will always speak to us. Approach the same object contracting your sense of taste. Do you have the same relationship to that object?

Now try the same thing with your sense of smell. First tell yourself that you have this sense of smell and you will know right away where it is located and what it is. This will make it possible to contract this sense of smell. Sustain the activity and follow the impulses. Say yes to what is happening to you, grab hold of it and let it take you. Notice how you deal with the world around you.

Now expand your sense of smell and sustain the activity. Say 'yes' and follow the expansion. Notice how your relationship to things has changed. You can do the exercise with all five of your senses and you will discover a rich, nuanced and suggestive world that is easily accessed. You will also find that this is just the beginning of an approach to what Chekhov called 'pure act-ing', which is acting without justification, yet full of psychology.

6 QUALITIES OF MOVEMENT

When we understand the value of working with archetypes, Michael Chekhov's technique makes a new kind of sense and a real order begins to appear.

Chekhov rightly suggested that *How* is the primary question artists need to ask. He recommends that we leave the question of why to a later time in the rehearsal. If we begin by asking the question *why*, we will be engaged on a level that is both dry and cold; the question has little room for the imagination. He said *why* is a scientist's question.

Who does what how? This is a good way to order the work. In the end it will become necessary to answer why, but he suggests that *how* will give us the answer to *why*. *Who* is obviously the character, this is creative work and much of it revolves around *how*. *What*, are the things that happen, most of that *what* is given by the author. The 'objective' remains for the actor to work out, and this is also *what*. The question of *how* opens up the creative world for us. This question answered is what allows us to come back again and again to see great plays like *Romeo and Juliet*, because it is about interpretation. Every production of this play is different from another, because of *how* it is interpreted.

We can use a simple example by posing this question: *Why* does the earth revolve around the sun? It is a question for scientists and they can answer it in this way:

> Any two masses exert gravitation forces on each other (Newton's Law of Universal Gravitation). Therefore the force of attraction between the sun and the earth is large enough to make the earth veer off from the straight line path that it would have otherwise followed by Newton's First Law and to make it follow an ellipsis.

There is little here to excite the imagination. It is interesting, yet it does not serve the artist. If we ask the question: *How* does the earth revolve around the sun?, we are immediately brought into a world of images and qualities. We see images of day and night, time, the passage of the year, spinning and tilting, etc. All of these images have some power to stir the artist into action. *How* it happens speaks directly to action and quality, corresponding to willing and feeling.

We can look at this question of *how*, in terms of acting, by approaching it through *quality*. The quality of a movement makes it unique. Because it is movement, it speaks directly to the actor. If you were to pick up an object, even a newspaper with *absolute care*, it will awaken within you a concern for the object that could communicate the circumstances and your feelings for the object. If you pick up the same object *roughly and quickly*, something else will be awakened, and another circumstance with its corresponding feelings will be communicated. If we believe that we act because we are actors, then mostly any quality will speak to us, especially how we move.

In Michael Chekhov's book, *To the Actor*, the first exercises presented are to feel that you have a body, and that the body moves. Immediately, there follow exercises on different ways to move. He offers four distinct qualities of movement and each of these qualities is an archetype. The genius of this is that by working on one of these qualities you are working on so many different qualities at the same time.

The names he has given them are Moulding, Flowing, Flying and Radiating. They very neatly correspond with the four elements of earth, water, air and fire, respectively. Chekhov's exercises here are very straightforward: make movements with this or that quality. A new sense of movement is immediately offered that puts the actor's sense of movement on alert, as it were, to become aware of what the experience of moving in this specific way awakens.

Moulding movements (earth) are resisted by the space, as if you were moving through a space made of wet clay. Like sculptors using the whole body, you mould and carve and prod your way through the (earth) space. The movement naturally becomes slow and heavy, forceful and determined, exact and economical. The will comes into play and the idea of form becomes corporeal.

Flowing movements (water) are not at all restricted by the space, but are led along by the space. The space is flowing like a river. You move from one thing seamlessly into the next, without pausing, without choosing – just to keep up with the flow. It can be fast or slow, heavy or light. Moving like this you will find ease and charm, pleasure, joy, conviction. Moving in this way, one can also experience helplessness.

Flying movements (air) are gone almost before we know they have happened. Off they go out into the space, never to return. The space moves so quickly – flying away, taking all form with it as it flies off this way, that way, up and down, all around. A panic, or confusion, a chaos, or even a miraculous connection, or a realization come from flying.

Radiating movements (fire) draw attention to themselves like beacons of light in the night. These movements light up the darkness of the space, they allow us to make contact with other things outside ourselves. We can move to illuminate the world. The movement is full of understanding and compassion. It is always attractive and at the same time empowering in a way that is not hard or violent. The movements go beyond the body with energetic rays. These movements have purpose and clarity.

These are the delineating aspects of the archetypes, which help us to separate them from each other. Many more things are possible. Earth is much more than wet clay; it is also sand, or gravel, or heavy stone. Each image delivers a new yet related movement experience. If we put a bit of air in the earth then it is easier to move through, or we could come to a complete halt and crack our way through the stone. The earth could mix with water to result in mud or fertile land. The mud could freeze or become baked with the heat of fire into a hard cracked or brittle medium to move through.

Water flows, but objects in it float lazily upwards, resting upon it, supported and influenced by the nature of the flow. Water can also rise in a flood or ebb away. Tiny streams trickle and they also swell with rain. The water can be supportive or crushing. Waves move with tremendous force taking things with them either to arrive or be swept away. They also crash violently on a beach with weight and force. There are currents and whirlpools, etc. Water can mean all fluids and each moves the way it moves, blood, oil, puss, honey, etc.

Fire consumes fuel and gives off light, heat and smoke. It can rage or explode, it can dance, and lick, and spit, and sputter. It is destructive and helpful. It repels and beckons. Fire can scorch, and dry, and flare up. It can glow low and secretly. It can flash. It can light the way and warm the night. It can cause you to open up to it or to run away from it. Fire is attractive and engrossing, romantic or cataclysmic.

The element of air is fast and thin. Home for kites and birds and swirling dust. The tempo is fast, light and airy. Things are done as the crow flies, directly, or carelessly, indefinitely. The chilling wind and the balmy breeze are found here: gusts and gales, puffs and whirlwinds, updrafts and gasses.

These elements are the building blocks of the universe. We can use them to build our own universe. They are the building blocks of our art, forming relationships, actions and ways of performing the character. They meet circumstances through quality and connect images to the reality of doing.

7 THE ARTISTIC FRAME: CONSCIOUS MOVEMENT

Normal everyday movement is generally without consciousness. We need to move to carry on a life, and our movements become more functional than expressive; we lose consciousness of them. As actors we have to make the connection between movement and expression, and between movement and life. The artistic frame is a device that we employ in the class, or for part of the rehearsal. We do not use it in performance because it takes too much attention, and our expression will appear stilted.

We acquire knowledge and facility by using the artistic frame. Every action of consequence has three parts to it, the preparation, the act itself, and the sustaining. These three parts are what make up the artistic frame. In exercising, it is critical that we use this tool. It is simple enough to do and well worth the effort it takes to make it happen. In terms of movement, it means that the movement will have three parts. The preparation is the engagement of the life-body, the act itself is the physical movement that is made, and the sustaining is the radiating of the movement. To radiate the movement means that we go on with it for as long as we can radiate it. Again, it is the life-body moving beyond the physical body. This practice puts intentional movement into the body. It makes many things possible and it is also pleasurable.

Exercise 8: The artistic frame

To experience the artistic frame, we can return to Exercise 1. Most of the artistic frame was already engaged in that exercise. Stand still and be present by feeling your feet on the floor. You will raise your arm so that it will be pointing towards the sky above you. Begin by raising the inner arm of the life-body (preparation). Then allow your physical arm to follow that preparatory movement (the act itself). When you have reached the end of the movement, when the arm cannot rise up any more, and you are pointing to the sky, continue to raise the inner arm of the life-body for as long as you can (sustaining). This is the artistic frame. Doing this makes the movement clear to you. It also trains the life-body to become consciously engaged. If the life-body knows how the movement is done, then it becomes possible to experience the movement without actually moving the physical body. This is required if you want to make a successful practice out of the psychological gesture.

8 ACTION: THE PSYCHOLOGICAL GESTURE

Perhaps the most important contribution Stanislavsky made to
the art of acting was his idea of the objective, and units of action.
It is a way to form the work and this helps the actor to sustain
the performance over time, because it is a solid footing, clear
and energetic. This dramatic action allows the actor to speak his
lines and interact with others in a way that is necessarily consistent
with the story or the conflict presented. Without it, the script
of the author would be merely written words spoken aloud.
Knowing how we are active in the scene is a real concern for
every actor. We can define it with words, verbs, strong verbs
and we can have these verbs in our minds and this will give us
a guide to stay on course with the intentions of the playwright.

We can also translate these verbs into archetypal statements
of action, which will lead us to gestures, and these gestures can
become our guideposts. Being in the body, these gestures (forms)
come to the actor directly as knowledge, or a physical connection
to the action. They can generate impulses to satisfy the action.
The impulses surge through the body, and this engenders a real
bidding to do. One doesn't have to convince oneself of anything,
one is not called upon to consider anything, because the intellect
is left out of the effort. The inner gesture is the spark to the fire
of life on stage.

Action has to be approached through the realm of the will
and this is centred low in the body. Unfortunately, student actors
are often led to it by way of the intellect and that is centred quite
high in the body. This thinking causes some difficulty, some
faltering and floundering. Action is not the character's thinking,
it is the will of the character taking on a form. Clearly action is
about doing and not about thinking. What am I doing? A question
we inevitably come to as actors; it leads us to the form.

What I am doing is very specific, the more specific the better,
but the gesture I seek for this action is alive for me when I can

find the essence of that action. If, for example, I determine that my action in the scene is to seduce the other, then I must find a gesture that is all about seduction. In seeking it, I will find that the gesture of seduction is a pulling in towards me. I am seducing so that I can have it. If I have it, then I have taken it. This is essentially what is going on, this business of taking in a very special way that is seduction. To take could be called an archetypal action and it holds the smaller actions of seducing, or spying, or plundering, or seizing, or killing, etc.

When learning about the objective, we have been led to look at it in this way: What do I want? This is helpful for the intellectual pursuit of finding it. For an actor playing Richard III, it might sound something like this: 'I want to be king.' This is okay, it has started to wake up something in the actor. In the end it will become more important to ask: 'How do I become king?' It is not so much any more about wanting something but about doing something. Richard becomes king by murdering, by stealing, by seducing, by seizing power. He is all the time taking in one form or another, with one quality or another. If the actor finds the gesture for this archetypal statement of action, 'I take', and works with it in many ways, it will take him far. The simplicity of the choice helps the actor to explore its various potentials and range. The exploration is through quality. To take slowly and sneakily is very different than to take explosively, which is different again from taking grandly. These qualities added to the gesture, supply the specific of each moment of taking, all the while the actor is involved with one simple gesture. There is no inner dialogue of doubt or consideration, no questioning of the self as to 'whether I am on the right track'. The gesture opens within the actor a steady stream of taking. Streams of taking are generating impulses that fulfil the action. The body comes alive in new and unexpected ways, and the actor engages us because he becomes fascinating. This is the real gift of the performing artist, to sustain a condition of fascination for the audience. As long as the actor is fascinating, he will engage the spectator in a complete way. We who work in the theatre are always fascinated by the potential of Shakespeare but rarely are we fascinated by the actor. Yes it is always our hope that we will be, but more often it is just the play that holds us – the language, the structure, the twists and turns of plot, the author's form. We are often let down by the actor, because he is bogged down with his lines and is living in his head and not his body.

When we look at action in an archetypal way, we find that there are not so many archetypal actions. Everything begins with wanting and then leads to something else more active and direct. 'I want', is itself an archetypal statement of action. There is a gesture that clearly speaks this, a primitive and lovely gesture that wakes up in us these streams of wanting; it may even be the very first gesture we make to the outer world as human beings. It is a gesture from the infant who sits alone and calls out to the mother, not with words but with the body. The gesture says, 'I want comfort, I want food, I want you.' As you read this, it is possible to see this gesture, because we all know it, we have all made it. And if you make this gesture right now, you can feel the streams of wanting moving through your body. It bids you into this action.

Many things we want and many things we do not want. The things we do not want we actively reject and here we find ourselves with another one of these archetypal statements, 'I reject'. This too is primitive, active and evocative. Again the child without language becomes our guide. We see the child sated or utterly discomforted making this gesture of refusal and rejection. The infant as guide is so clear, because there is no language except gesture to communicate the most primitive things. As we become more sophisticated, our primitive needs and wishes do not disappear from us. They stay within the body and we are in a direct yet unconscious contact with them. We now have words and concepts, ideas and proofs of 'why this' and 'why that'. We easily confuse one thing with another, so many things have names now and we must remember them all. When we come back to the archetypal, then things are simpler but no less profound.

There are times when we cannot be persuaded or we will not be persuaded. Our opinion, our point of view, our stand must hold and nothing will change it. 'I hold my ground' leads us to a fine gesture where we become rooted in our beliefs, we dare another to change things even though we know that it will be impossible.

Sometimes, after long argument and conflict, when I can no longer hold my ground, after I have been beaten or persuaded, then it becomes necessary to yield. This can run the gamut from completely submissive to a very arrogant or reluctant yielding. Another archetypal gesture can be employed here from the statement 'I yield'.

Generosity within us is a powerful force. It is so difficult to watch another suffer even slightly. We do what we can to help them. We joke with them, pray for them, cheer them up, kiss them, console them, slap them out of it, or challenge them to help themselves. All of these actions and more live in the archetypal statement of 'I give'.

This work has enthralled me for some time and I have examined it with scrutiny. What I have discovered is this. There are six statements of action. These statements could be called archetypal, and all other actions or objectives are based in: I Want − I Reject, I Give − I Take, I Hold My Ground − I Yield. Try as I might, I cannot come up with others. I found that these suffice. Because they are archetypes they hold so many things within them. Qualities are practically infinite in number, and quality will always change the archetypal to the specific. Kissing and punching, which seem to be opposite actions, are truly both giving. One of them is tender and soft, the other is violent and hard. The specific gestures themselves may differ as well, but essentially it is something coming from me and going to you.

Again it is best to be as specific as one can be concerning the chosen action. It will not do to simply say, 'I am giving' if what I intend to do is to cheer you up. This clear choice needs to be settled first. Yes, I am cheering you up. The next question is HOW can this happen? If I begin to talk out loud about how I can cheer you up, and while I am talking about it I also use my hands to help me speak, I will find that I unconsciously begin to make gestures which are very much about giving. Now I know that this is how it is done, and I can find the psychological gesture of giving with its light quality and upward direction. My mind is satisfied, so I no longer have to think about it. But better than that, this gesture of giving begins to wake up impulses in the body. These impulses help me to lift you out of your doldrums or cheer you up.

In training we work with five gestures as archetypes, and for training purposes these five are rich. Pushing, pulling, lifting, throwing and tearing are a means of realizing the six statements of action. There are six directions to exercise them in: forwards, backwards, up, down, left and right. There is different information from each of these directions, and there is, as stated above, an infinite number of qualities to work with. Qualities are merely adverbs. Of course this can become tricky business in the beginning, so students are cautioned to work with qualities that can easily be imagined as a way to move. Words like tenderly, slowly, quickly, lightly, heavily, quietly, carefully, carelessly, sneakily, explosively, sluggishly are highly recommended. It is best in the beginning to avoid emotional adverbs, because actors can fool themselves into believing they are moving angrily, for example, when rather they have become angry and start moving. The former is full of artistic potential, while the latter can become a hazard for the other actors on stage.

Exercise 9: Psychological gesture for the objective

Say, 'I want' and repeat these two words again and again until you feel a gesture forming within you. Then stop saying it and just begin to execute the gesture. Make it large and full-bodied, put as much of your consciousness into it as you can. You must truly know you are doing this. Feel what is going on within you while you are making this gesture, especially when you radiate it. The gesture wakes up *streams of wanting* within you. Recognize the impulses that come from the streams of wanting that are being generated by the gesture. Use the artistic frame now, so that you can learn to make the gesture without the physical body. The artistic frame is invaluable at this point, it will deliver to you the real meaning of your gesture. Using the artistic frame now will allow you to bring this gesture onto the stage later if you need it. It will help to make the gesture an inner gesture. It will make the gesture one of the 'intangible means of expression'.

Repeat this process with the remaining five Archetypal Statements of Action. Speak the words until you feel a gesture forming within you. Stop speaking the words, use the physical body and make the gesture. A full-bodied gesture is what you want. The gesture should take up a lot of space. There is nothing mundane about it. These gestures are primitive; they are effective because they are primitive. You will feel it by speaking the words. The gesture will wake up the will. Use the artistic frame. Spy back, tell yourself what you experienced, reinforce it all with the spy back.

9 THE SWEET SPOT: SUSTAINING AN INNER MOVEMENT

In any action there is a moment that could be called the high point. It is when we receive the most satisfaction from the action. If I really wanted to slap someone in the face, I would use a fairly large gesture to do it. If I followed through with it, I believe the most satisfying moment would be when my hand made contact with the face. This could be called the high point, or better, the Sweet Spot.

When we work with the life-body, we are able to enter into a fantasy time/space where it becomes possible to sustain a very little movement for a long time. In the case of the slap, it would be a matter of continually living in the moment when my hand makes contact with the face, *that moment is happening continually*, not over and over again, but always, it is always new and fresh, the hand contacting the face. That is the sweet spot and I can sustain that. Although we have made a very large physical gesture all we need from it, in the end, is a very small piece. When making a large gesture and using the whole body, we become physically committed to it. It is so complete that we put ourselves in a position to become excited by it. We need not feel doomed to repeat this large inner gesture over and over again. Instead, as we further explore and investigate the gesture, we will find the sweet spot; the place in the gesture where we recognize we are getting the strongest excitation. It is much easier to do this than to talk about it. If you can already move the gesture inwardly, then you can sustain the sweet spot. This kind of conscious effort needs to be explored and used in the rehearsal. If you do that then, the body will remember it all, and be able to do it without effort in the performance.

Exercise 10: Sustaining the sweet spot

Create a psychological gesture that expresses the archetypal statement of 'I take'. Work with it, using all of the body. Feel satisfied that the gesture is waking up streams of taking within you. Use the artistic frame so that your life-body knows the gesture and also knows how to make it without the physical body. Try to sustain the gesture inwardly for as long as you can, follow the impulses, etc. When you are satisfied that the gesture is a good one, i.e. that it has the power to move you, then you can begin to look for the sweet spot. Return to making the gesture with the physical body. Pay close attention while you are making the gesture, as you are looking for the moment when you get the biggest kick or excitation from this gesture. Be clear about the physical location of the sweet spot. Now use the artistic frame again, so that you can introduce this sweet spot to the life-body. Now use only the life-body and live in that moment always. You do not have to repeat it over and over, but you only have to *experience* that moment as continually happening. You are engaged with a sustaining effort here. It is an inconceivable occurrence, yet it is quite easily done. It is one of the *intangibles*; you cannot put your finger on what it is, but clearly you will experience something. It could happen for the whole scene if you needed it to, or it could last for a monologue, or a brief moment.

10 SENSATION: FLOATING, BALANCING, FALLING

We have all touched fire before and we have been burned by it, so we do not have to think about how to behave when handling it. Every experience we have ever had has been witnessed and felt by the body. The body has reacted to it, and the body has also recorded it. If we seek the experience as a sensation which is something felt by the body, as opposed to a memory of a specific event, then we begin to discover that these sensations are tied together, all of our sorrows having been forgotten by our conscious selves, but never forgotten by the body are alive as sensations. The specifics of any past events are no longer important to us. What we find is that the response to the event is there within us. It can be felt again, triggered by the inner movements we made when we first experienced this event. We also discover that all the sensations that produced a particular reaction in us live all together in what Chekhov called our subconscious laboratory. All joys, all fears, all jealousies, all regrets, all loves, all pleasures, all doubts, all sorrows, all hopes are there and all this is known by the body. They have accumulated, each in his own house, each house living within us as archetypes. I, the actor, can summon them.

A curious thing about Michael Chekhov's investigations is that they invariably yield a limited choice of possibilities. But soon we begin to understand that with the proper application these few choices can exponentially grow into a multitude. Clear examples of what I am talking about are these sensations. Just before his death, Chekhov began to experiment with sensations as archetypes. He discovered that there are three primary archetypal sensations. The first sensation, *floating*, holds all the positive feelings we might experience. This physical sensation is essentially our ability to move upwards. It is revealed in our language in more idioms than one, but one will suffice here. We speak of our spirits being lifted. Sensations of joy, pride, love, freedom, hope, etc. move in an upward direction, and we experience them as a kind of floating up. The second sensation, *falling*, holds all negative feelings we might experience. We speak of being down in the dumps, falling into despair, etc. Sensations of sorrow, doubt, confusion, panic, despair move in a downward direction and we experience them as some kind of fall. The third archetypal sensation, *balancing*, or seeking the equilibrium, holds the transitory sensations of understanding and revelation. These moments of balance are when we gather all of our forces to keep our feet on the ground, as it were, to not fall, not to float away. It is so easy to fall, so easy to float off, but it requires much work to stay balanced and awake and this has its accompanying sensations of calm, collected reserve, power, sobriety, etc.

Obviously we cannot literally float away. We have gravity to keep us on the earth, nor are we continually falling down just to get up again. We are actually in a kind of numb physical balance. These actions of floating and falling can be looked at as purely psychological things and can be translated into physical understandings or in this case physical sensations. They become very dynamic realities for the actor to employ. The body feels the sensations. Within the body they correspond to inner movements, or movement impulses. Gymnasts and acrobats can learn to fall gracefully and effortlessly without any panic; this is their work and we applaud them for this ability, but any normal person who has the slightest bit of a fall, or even near fall, will receive an instant and definite panic that is experienced in the pit of the stomach. Just recall a time when you went to sit down and judged the chair to be a certain distance beneath you, then you give into the sitting, only to discover that you had miscalculated that distance by just a few centimetres. The resulting sensation in the pit of your stomach is enough to cause you to let out a cry or a gasp of fear, which always results in a little laugh once your buttocks find the chair and your equilibrium is restored. This fall of a few centimetres is a dynamic experience; it is in fact a real event that can be useful to the actor. There is also the moment of desperation when we are awakened from a dream in which we have begun to fall. This is very primitive business, but nonetheless very human. It all ends, however, once the equilibrium is restored.

So the work becomes: How can I sustain a psychological fall? How can I sustain a real panic that wants to resolve itself in balance? Michael Chekhov's approach always comes back to the imagination. So to speak about a sustained fall here, we must look at it as an imaginative fall. A fall that begins in the imagination but is felt, and doesn't end until the actor ends it. It isn't actually the fall that interests us, but *the activity of falling*. When the human body is falling, there is an accompanying sensation. Chekhov said that the door to feelings is opened through sensation. The process is a clear one. We know as actors that we cannot appeal to the emotions, because we risk coming up with nothing but tension. We hope for the best and trust in inspiration. It is through our feelings that we communicate; Chekhov said the feelings are the language of the actor.

Actors come to believe that if they think sad thoughts they will become sad. But what is in fact happening to us as humans is that we are thinking sad thoughts because we are sad. And that it is our bodies and the sensation of sadness within the body that is leading us to have sad thoughts. We fail to notice that we are sad in our hands and shoulders and legs, that our movements are heavy, and that we are having sensations that are downward moving. This is always true and we can recreate these downward movements with our imaginations. Once the sensation begins, the natural flow of events comes unimpeded, so that the sensation awakens the feeling and the feelings lead us to the emotion, which is the final outer expression seen by the audience.

It follows then that the opposite is true of upward movements and their accompanying sensations. In the imagination, it is possible to float up, to sustain this floating up and to experience the body or parts of it moving in an upward direction. The sensations that follow are ones of pleasure, joy, victory or freedom.

The balancing sensation is a bit more elusive, because we take for granted our perpetual state of equilibrium, and only experience the sensation of seeking the equilibrium in order to prevent a fall. In training we bring ourselves to the point of falling whereupon we catch ourselves almost as if we were tightrope walkers who must use all our powers to stop ourselves from plunging to our death. This is a very powerful sensation, a moment of revelation and strength. With practice this sensation can be sustained and we can prolong the feeling and use it as we need it. These three primary sensations is work on the vertical line.

The horizontal line, with its directions forwards and backwards, is equally powerful. The sensation of fear is backward moving, a retreat or flight mechanism that is quite easy to engage, and produces a curious effect of doubt, timidity, apprehension, concern, etc.

The forward moving sensation is one of a very active and sure will, confident, expectant, assured, resolute, etc.

The directions of purely right and purely left are subtle in their psychological meanings. If, however, we look at these directions as working simultaneously, then something very interesting begins to occur. We can experience ourselves either growing or shrinking. This expansion and contraction is full of possibilities, and is at the very bottom of Chekhov's technique. It is possible to form these principles into gestures, or to experience them as sensations, also as inner movements.

Chekhov's technique is always striving to lead the actor to an objective understanding of the human condition. These things spoken of here are universally human, they belong to all of us, and when we contact them we have an immediate affinity for them because we recognize them. But more importantly, when the audience watches an actor engaged with physical sensations, they experience a sympathetic response. The audience declare they were moved by the performance, because something in them was in fact moving. If they took the time to analyse what was going on, they would find that the idiom they just used was simply true.

Exercise 11: Experiencing sensation as inner movement in a direction

This exercise is one of the few included in this book that comes directly from Michael Chekhov. He says that sensation is the simplest and clearest way to access the feelings we need to express. By appealing to the sensation, which is a physical thing, we are on solid ground because our orientation is always with the body. Say to yourself, 'I want to experience the sensation of defeat.' You can trust that the body knows this sensation. Any time in your life when you experienced defeat, your body was with you, and your body recorded it as a sensation. If you give yourself the time and space to experience this sensation after you ask for it, you will experience it. It is simply the ability to listen for a movement and recognize its direction. You will feel this sensation of defeat as an inner movement that is moving downwards. It is pulling you down, so to speak. This is the physical sensation and this now awakens a *feeling*, and this *feeling* will give way to the *emotion*. The emotion is the outer *expression* that is seen by the audience. It is quite fantastic that you can produce this sensation by commanding it, even if it is not connected to any circumstance, or any *why*. Simply by asking for it you can have it if you remain sensitive to what is going on with your body. You can repeat this simple command, 'I want to experience the sensation of', and try various sensations. The body knows them all. Here are some examples: 'I want to experience the sensation of', love, fear, shame, power, victory, freedom, sorrow, joy, standing in the moonlight, doubt, jealousy, grief, pleasure. Ask for them and you will receive them. You want to experience the *physical sensation*. Do not worry about the emotional aspect of it. If the emotion presents itself then go with it, but just live in the sensation. The sensations are strong and they will lead you. What is most interesting here is that it is all somewhat abstract, without justification, circumstance, or cause. It is your choice as an actor to have and use these things. When the imaginary circumstances of the play are around you, and the *character* is involved with them, then the *character* will receive the sensation that is correct for that specific moment. These are simply sparks that will enflame the necessary emotional life. If you are an actor, then the circumstances alone ought to be enough to ignite you. If they do not, then you can find what you want in this way.

11 CHARACTERIZATION: STICK, BALL, VEIL

To create something good, it is best to have a clear foundation to build upon. It is a very simple thing to read the play and then make a decision based on the first reading. In doing this we have already begun to probe into things, there is no need to analyse anything yet. One idea and one image can take us into the play from the point of view of the character. We could call it a first bold stroke, and it can open the door to the mystery of the play. To look at Hamlet as an example, by merely reading the play, he appears to us as a thinker, his thoughts trouble his soul, and his struggle into action is the main conflict he wrestles with. This simply makes sense and is completely justifiable. We can feel comfortable with this beginning. The picture of him is complete as a human being, and we have a container in which to place the play. This view of Hamlet as: 1. thinking, 2. feeling, 3. doing character leads us to images that can evoke within us real creativity. We can call him a thinker, because this is the first of the three functions he engages. This is an important distinction and delineation in our search for the character.

12 THINKING

Thought as a function moves with a particular quality. It is direct and it works to find its way to the target piercing through non essential things, separating what is true from what is false, what will work from what will not work, etc., much like an arrow moving swiftly through the air. This arrow is an apt metaphor because its structure is that of a stick. Thinking is a linear process and the stick is an image that is ripe with possibility for us. If we apply it to the body, interesting things start to happen instantly to our psychology.

Exercise 12: Incorporating an image, a stick

Picture a stick, and now imagine that your whole body is a stick and the stick can move. To begin with, you must become a stick; completely forget that you are a human being. This is peculiar.

At first it is a total body investigation of the image. You are educating your body to know how it is to move like a stick. You will find that your movement becomes stiff and very rigid. Try to move in as large a way as possible, taking up as much space as possible. Do this for a while until you feel that you have truly incorporated the image, and that the body understands it. Once this happens, it is now possible to allow the rigid movements made by the body to soften on the outside; it is time to shift your concentration from outer to inner. Let the stiffness fall from the physical body but continue to move as a stick with the life-body. Do various activities that are real and practical, and pay attention to what occurs with your psychology. Ask yourself if this is how you normally experience yourself, and then try to identify exactly what you are experiencing. If you are working in the right way, you will begin to notice a shift in being. This is not a character per se, what you have done is opened the door into what could be called the 'House of Stick'. Continue to do activities that are real, walk, sit, stand, lie down, pick up an object, handle it, give it to another person, look at things, touch objects, etc. Perhaps you notice a concentration of energy somewhere in the body. Hopefully you will feel different from your normal self, because you are moving with consciousness, and this is something we do not normally do. To move with consciousness is very important. The image comes with information for the body to absorb. It is quick and simple.

Now that you have created and entered the House of Stick, it is possible to visit the various rooms within it. You can become specific about the type or quality of stick you wish to incorporate. Let's say the stick is now a toothpick. This is a very special kind of stick, it is pointed at both ends, very thin and fragile. One could pierce and poke with it, but it remains thin and is easily broken. Using the toothpick as the image, return to the concentration on inner movement. You will find that it is much more specific and it gets closer to character. However, there is not yet a complete character in your body. Other possible rooms that could be in this house are baseball bat, beautiful inlaid wooden chopsticks, police truncheon, iron pipe, pencil, tree trunk, sword, hatpin, etc. Any object that is rigid and moves in a stiff and unbending way could belong in this 'house'.

The type of stick that belongs to Hamlet is of course up to you, and how you are able to imagine Hamlet or the qualities which belong to him. Bringing this new way of moving and behaving into the rehearsal will lead you to make choices that come out of your intuition and association and imagination, as opposed to coming from the rational mind. With a very simple image it becomes possible to begin without a plan. You will find many things for the character through this and as you proceed with your rehearsals, other finer images will offer themselves to you, almost as if you are a magnet and this initial, primitive work will become covered by richer and more sophisticated things.

13 WILLING

If you were to work on another character that is not a thinker but a doer or willing person (let us say Romeo, whose first impulse is to do things), then you would need to visit another house. If the three functions are put into their proper order for Romeo we see that he first takes action, then he has feelings, and after that he thinks about his feelings. The image to explore here is that of a ball. The ball that rolls and bounces, and would if it could continue to stay in motion, is an appropriate image for Romeo. When we imagine something like perpetual motion, we are led to the image of a sphere, like a planet in our solar system.

Exercise 13: Incorporating an image, a ball

Again begin with the picture of a ball. Then make a full body exploration so that you can incorporate the ball. It is not necessary to get down on the floor and roll around, although you can if you wish. Rather try to work with the shape and the feeling of rolling and then bouncing. Roll into things, bounce off them and continue in a constant motion of doing. You will find that your movements are in fact continuous and it is almost as if you are constantly looking for something to take up your attention in an active and spontaneous way. Continue like this until you feel as if your body has understood the essence of the ball. Now shift your attention to moving inwardly like a ball. Let all the abstract kind of movement slip away from your physical body and begin to move outwardly as normally as is possible. You are a human being now, yet the energy of the rolling, bouncing ball is leading you from the life-body. Ask yourself if this is anything like your normal experience of yourself. If it is not, how is it different? Begin to do various activities like you did with the stick. Involve yourself in activities much like the character would do. It ought to become clear to you that this is nothing like the stick. Still, it is not a character. This is only a way to lead you there, you are now in the 'House of Ball.' If you are only exercising this new way of looking at things, then it is important to create and visit as many of the rooms of this house as you can. Some possible rooms are ping pong ball, football, billiard ball, beach ball, medicine ball, egg, basketball, baseball, etc. What connects them to the House of Ball is that these objects roll and bounce into and off things. Each of these balls is quite different from the others, each has a particular way of moving, each has a different purpose, and each has its own quality. Each of these balls can express a different kind of will, and your investigation, as the imaginative actor is to find the ball that's fit for Romeo. We can call Romeo a doer because this is his first function, it is not as if he only wills himself into action, the play would not live without his feelings and his thoughts. It is a very special way of being we are looking for, and a way to transform from our everyday selves into a character that is consistent. He is a willing type of man and the ball can help to contain him within this for you.

14 FEELING

The next type of person is a feeling person. Juliet could be seen as one. Her feelings lead her into action and the action is then finally considered. This type of character needs another image to concentrate on, and this image is a Veil. This object requires outside influences to move it. The veil is a light translucent piece of cloth, its movement is initiated in one place and the rest of it follows. It stands inert until a wind blows, or it is picked up and thrown or dropped, it is delicate and at the same time strong. It is soft and its way of moving is flowing and indirect.

Exercise 14: Incorporating an image, a veil

Repeat once again the procedure of a full body investigation of this new object. Allow yourself to become a veil through your movement. Let go of the human for a while and just move as the object. Notice how you are moving now, and that the movements are flowing, soft, light, easy, yielding, quiet. Keep it up until your body has absorbed the information from this object. Now begin to concentrate on the life-body moving as a veil and the outer body becoming human. Outwardly the movement is no longer peculiar, there is nothing that draws attention to it but inwardly the image is very much there. Once you have done this, begin to do real things, work with objects and relationships to objects. All the while allow your inner body to lead you. Ask yourself the same questions you did with the stick and the ball. Is this how you normally experience yourself? If not what is the new experience? Is there a concentration of energy anywhere in the body? Have you transformed? Realize that you have entered the 'House of Veil' and how different it is from the stick or the ball. Continue to exercise so that you can create and visit various rooms in this house of veil. As with the others, you are looking for objects that can connect to the house. These objects have a soft quality in the way they move, yielding and flexible. Possible objects that belong to this house are cotton string, iron chain, a golden thread, a rope, a curtain, a delicate piece of antique silk, a leather belt, a wet towel, a blanket, etc.

When you move like this you come to understand certain things. One thing you will immediately know is that Juliet cannot move like a stick. It is practically impossible to imagine her as a stick, but you could be comfortable with her moving as a veil. Which veil belongs to Juliet is entirely up to you.

The purpose of this work is realized when you make a psychological identification with the chosen image. You soon begin to realize that every image has a psychology, and every image can lead you to something that exists outside your everyday life, but at the same time is familiar, the world starts to open up to you as a source for your creativity.

These three archetypal images lead us directly to types. The three types illuminated here are truly living in the world of drama. The type can contain the characters we work on, but we cannot rest satisfied with just a type. We need to go deeper into things from the perspective of the type.

15 CHARACTERIZATION: ARCHETYPE, THE PSYCHOLOGICAL GESTURE

The dictionary defines archetype as the prototype, the type from which all types derive. One could say that it is an encompassing image of something. It also contains smaller ideas revolving within it. Take for example the idea of the cat as an archetype. It is easy to see that a lion, a tiger, a leopard, and a lynx are each different animals, but they all are cats. The archetype of cat is able to hold all of them collectively, while not diminishing the fact that each is at the same time individual. If we were studying these animals, it would make our work a bit easier to view them first as cats then as lions, tigers, leopards, and lynxes.

The pioneering psychologist Carl Jung had very much to say about the impact that archetypes as collective images have on the human psyche. His work and the work of his followers is dense and illuminating. Suffice it to say that these specific images have found their way into the lives of human beings across different cultures. The images reside within us in a place Jung has named the collective unconscious. Cultural history has poured itself into this collective unconscious, it is a region within the human psyche that is active, yet hardly in our conscious control.

The ideas about acting developed by Michael Chekhov rely very heavily on this idea of collective energies. We find, through exercise and practice, that we can expect specific responses to certain images. If a room full of actors is asked to create a large movement of the physical body that could express the archetype of the hero clearly and succinctly, we would see that virtually everyone in the room will move in the same direction. Heroes are everywhere in history, in all the great literature, from David slaying Goliath, to Luke Skywalker defeating the evil Empire. This image lives in us, and we do respond to it by moving the body. What is of particular importance to us is the *direction* that the body is compelled to move in. In the room of moving actors, we will see that the actual gestures created by the actors will differ from each other. This is the result of the individual actor making the movement; but all of these movements will be in a forward and upward direction, because this is a collective response to the energy of this archetype. The *direction the movement wants to go* is the useful information for the actor, because he can rely on it as a living truth. It is first, an impulse. If you move in this direction, there is information for you to take and use. If I witness the movement as a spectator, I understand something about what is going on. This understanding is not conscious; it is felt.

Connections are made to larger trans-personal ideas as source material to create with. The archetypes are how the unconscious can communicate to the conscious, and the body is the medium of this communication. We can also take the process in reverse. By making a psychological gesture that corresponds to an archetype, we can touch the vibration within the unconscious resulting in an excitation of the conscious. This is essentially the Michael Chekhov acting technique. In Chekhov's own words:

> All you experience in the course of your life, all you observe and think, all that makes you happy or unhappy, all your regrets or satisfactions, all your love or hate, all you long for or avoid, all your achievements and failures, all you brought with you into this life at birth, your temperament, abilities, inclinations, etc., all are part of the region of your subconscious depths. There, being forgotten by you, or never known to you, they undergo the process of being purified of all egotism. They become feelings *per se*. Thus purged and transformed, they become part of the material from which your individuality creates the psychology, the illusory 'soul' of the character.
>
> (Michael Chekhov: *To the Actor*)

To work on movement in this way has a few benefits: It leads the actor towards making very defined and delineated movements which are aesthetically pleasing to watch and to execute, it also encourages a feeling for form, but most importantly it trains the actor to move what Chekhov calls the inner gesture. The psychological gesture must in the end become an *inner gesture*. It is found with the physical body, it corresponds to the archetype and is archetypal in its form. This gesture is never shown to the public. It *must become an inner gesture*, an archetypal image that is in Chekhov's words, 'a crystallization of the will forces of the character'. This is another application of the psychological gesture. The psychological gestures of action say 'I do this now'. The gesture of the archetype says 'I am'. The gesture helps us know what this special quality of will is. It is the character who does things. 'I am doing this now'.

To find the correct archetype as a model for the character is very simple. Aristotle said that a man is the sum total of his actions. You must read the play and make a list of the deeds done by the character in the course of the play. It is through what has been accomplished that we can understand an individual. Just stick to the facts given by the author within the finite world of the play. You can call this the deeds done list. When you have the list of these deeds, you will be able to draw a defining conclusion about the character. The archetype is the thread connecting these deeds one to the other.

The image alone is helpful, but you can get deeper into the will if you make a gesture of the archetype. If you physically honour the impulse by giving it a form, then the image will be incorporated. The psychological gesture is the tool that will fix this *quality of will* in your body. The talented actor concentrating on this particular tool begins to make connections with other tools that have been engaged during the rehearsals. It's a matter of applied energy travelling on different circuits, each vibrating in sympathy with one source. Using archetypes as dynamic vibrating energies, our task is to set up a condition within ourselves so that we can have sympathetic vibrations to them. These are honestly felt things by the actor, real food for artistic self-expression. This is how the Fourth Guiding Principle is put into play.

The actor does not enter the stage screaming the archetype; rather the character feeds upon it and easily reflects it in all his behaviour. Chekhov did not recommend that the actor present the archetype as the character. The image has too much power; it is not a clearly defined individual. Actors presenting only archetypes in their performance appear strong, but general. They are a bit blurry and quickly lose interest for us because nothing can unfold. It is merely force thrown out. It can be astonishing for a short duration, and can perhaps be useful in a stylistic or formalistic fashion. It is not the thing itself that interests us, but the type of will force it has. Consciously or unconsciously this is what the author has built the character upon, it is the energy behind the sum total of his deeds. This approach to seizing the essence of the character is direct. It is, as Chekhov says, 'the first clear bell we ring for the character'. Through rehearsal the energy of the image becomes known directly into the body, because the body creates the psychological gesture for the archetype, thereby experiencing directly the vibration of its energy.

Engaging with the archetype gives the creative individuality of the actor something it can sink its teeth into. Each actor will respond to her own image, and each actor will know when she has arrived at the image that will serve her. Some clear confirmation will present itself as if to say 'this is the image to work with'. A kind of bell will sound within. The real purpose of working with the archetype is to find a synthesis of all the disparate elements before us. Something must hold it all together, one guiding principle, one feeling of the whole that makes it possible to act.

Exercise 15: Direction and the psychological gesture

To find the psychological gesture that expresses the archetype, begin by standing still and quiet, feeling both feet on the floor. Softly name an archetype, forming the word and speaking it. Put your attention inside your body. After saying the name, you will feel an impulse to move. If you are waiting for it, you will feel it. This impulse will be direction driven. You will feel like you want to move up or down, forwards or backwards, expand or contract. You also may want to move in combinations forwards/ up or forwards/down, etc. If you are oriented in the six directions during the exercise, you will understand and experience what you are looking for. Try this a number of times using different archetypes. Stand still and quiet, speak the name of the archetype, wait for the impulse and recognize which direction the impulse wants to go. Here are a few possible choices: the king, the fool, the loser, the mother, the hero, the slave, the warrior, the victim, God, the Devil, the whore, the thief, the orphan, the father, the gambler, the hermit, the outsider, the soldier, the dreamer.

The psychological gesture is a large full-bodied movement that expresses the essence of the archetype. This gesture must move. Because you know which direction the archetype moves in, you know 85 per cent of the gesture. The remaining 15 per cent is to give form to this impulse using your arms, head, legs, hands, feet and torso. You already know the gesture. So, 1–2–3 go, just make it. The doing of it will tell you if it is good or useful to you. If it speaks to you, and you feel the energy of the archetype, then you have a gesture you can develop. If not, then abandon it and find another. When you apply yourself to this activity, you will quickly find your way into a rich world of sensations, images and impulses. You can develop the gesture in numerous ways. You can change the quality of the movement and you will receive a nuanced experience. You can change the placement of your weight and it will be different again. Any little modification you make to this gesture will give you another shade of the archetype. It is possible to find a very particular quality of will by developing the gesture.

16 CHARACTERIZATION: THE IMAGINARY CENTRE

It is more than obvious that when we are thinking we are using our brain. This thinking part has a very specific location in the body. We can comfortably say that the head is the thinking centre. Here we calculate, plot, scheme, dream, consider, determine, analyse, rationalize, ponder, invent, accept and dismiss. With all of these we are active in the head and if we can say that Hamlet is a thinking type then it would follow that he is initially active in the head. It could be said that his life is centred there and from there he proceeds with all the rest.

Juliet, being a feeling person initializes her life from a different place. Although it is a poetic notion, everyone accepts that the heart is the seat of feelings. Hearts break and heal, they soar and sag, they can be warm or cold, they can beat quickly in fear and beat easily for joy. An open heart is always a pleasure to encounter. A closed heart is so difficult to be with. We could say that the heart located in the chest is our feeling centre. Juliet's life begins there and from there she proceeds with all the rest.

Our will, which is located low in the body, is a result of our appetites and desires. The entrails, groin and upper thighs are all ignited by our desires. We could say that the pelvis is the willing centre. It is from here that Romeo's life begins, and from there he proceeds with all the rest.

When we start to observe the physical life of people, we notice these things. Just by watching someone walk, it is possible and quite easy to see where their centre is. This centre is a centre, because everything comes from it and returns to it, it seems to hold the organism together in an efficient and comfortable way for the type.

We already know the type, because the author has indicated it to us by writing the character in a very specific way. We have explored the type by incorporating an appropriate image with our physical body. We can experience the world of the character now and make our way consistently towards something particular which is held together by the type. The imaginary centre is a way to help us there. With it we locate the precise place in the body from which we will move.

This is another variable that could be brought into the psychological gesture for the archetype. Moving from a centre has a powerful impact on how you will move the body. This is a building block in developing the psychological gesture for the archetype. The imaginary centre is also a completely effective tool in itself to define your character. It is a sure and confident way towards transformation.

Exercise 16: Moving from the imaginary centre

Stand with both feet firmly on the ground. Lift your arm up as if to wave goodbye. Move with consciousness, know that you are moving the arm and try to experience the movement as if for the first time. This will allow you to know what it is you are doing, or what it takes to move your arm. Familiar movements like this are almost always done with no consciousness. After we know what it is to move, then we can move in a very particular way and have new and rewarding experiences from our movements. Notice that this movement of the arm seems to begin in the shoulder or upper arm and it continues to its completion.

Now *imagine* that your arm is connected directly to you head and it is possible to move the arm from the head. Of course the arm is not directly connected to the head. If you experience your body as a unit, as one incredible thing in which all the parts are connected to all the parts, then you can do this. This connection is an imaginative and energetic one. It is not muscular. Once we try it, we understand immediately that it is possible to move the arm from the head. Notice how your sense of waving goodbye has shifted somewhat from your normal way of doing it. Try it again with the other arm, and then try to do different kinds of movements like touching an object, or walking, or sitting, or even speaking. Allow yourself to do everything as if the finger, the buttocks, the feet, or the voice were connected directly to the head.

Now try to wave goodbye as if the arm were connected directly to the centre of the chest. This I am sure you will notice is quite a bit different from the previous movement, which was connected to the head. Try many different things, consciously beginning your movement from the centre of the chest. It is good to repeat the same activities you did when you moved from the head centre, just to see what the differences are in terms of your sense of self.

Once you can really feel the difference, then begin to repeat the same activities, but now the movement begins in your pelvis and all the parts of your body, from your little finger to your lips, are connected directly to the pelvis.

Moving from a centre produces a continuity of being, everything we do begins to harmonize around this centre, it allows us to find our way to a specific kind of behaviour and we can return again and again to this same sense of being, because the psychology is always reflected in the body.

Putting your attention on the imaginary centre is a very clear path to defining the physicality of the character. To be able to move from the head, or the chest, or the pelvis is just the beginning. At this point it is all simply mechanical. Once we begin to engage with images, then real transformation occurs.

Exercise 17: Defining the imaginary centre with an image

Now that you can move from a centre, begin to experiment with different images within the centre. Imagine that in your chest is the sun, and that you can feel the warmth and the power of this sun radiating from the chest up into the head, and down from the chest through the trunk into the legs and feet, and across the shoulders and arms and into the fingers. The sun in your chest is your centre and it touches every part of you. This or any image you choose to place there will cause some kind of energy shift within you, perhaps your breathing will feel different, or the way you make contact with the floor will feel altered. Just note the shift. Now move the arm as you did in the previous exercise to wave. But allow the sun, which is in your chest, to move the arm for you. Give yourself over to it completely and trust that the sun can, and will, move your arm. Try other simple movements, but always allow the sun in your chest to do the work for you. Something, which may surprise you, will come of it. There is a freedom and a pleasure that comes from surrendering to the image. We give up the responsibility to something that is not our normal selves, and we open up to the glorious possibility of transformation. Let simple gestures give way to more complicated things like walking, talking, sitting, standing, running, etc. It is always the sun doing it.

This exercise can continue as long as you are getting pleasure from it. This way of transforming is so easy and free. You can change images and change locations. The imaginary centre can be anywhere you choose. Change the image of the sun to its polarity, a block of ice. Place the ice in the head and move, allowing the ice to move you, then place it in the chest, and then again in the pelvis. The image and the location work together to lead you directly to a clear psychology, but you have not been thinking at all about psychology.

These locations of head, chest and pelvis are not the limit of your choices. The imaginary centre can be anywhere you choose to put it. I suggest placing the centre in the head, chest and pelvis to begin with, because of the obvious connection to thinking, feeling and willing. It is even possible to have the centre outside of the body, a bit above the head, or behind the back, or in front of the chest, etc.

17 CHARACTERIZATION: IMAGINARY BODY

Proceeding from the standpoint that the psychology and the body are one thing, we easily discover that the type of body a person has usually determines a great deal about his personality. The form that the human body takes is always the same, but the size and proportion varies from person to person. The differences help to form something specific about our personalities.

When we read the plays of George Bernard Shaw, we get very vivid pictures of his characters because he describes them to us in detail, but when we read Shakespeare we get very little of this, so we invent them as we see them. Chekhov suggests that we read the play and first try to imagine what the character looks like, and how the play would be acted. When we read the play imaginatively, we see the characters going through the actions. Something happens when we imagine the character in the play, as a character, separate from ourselves. Doing that simple thing sets up for us a fantasy that surpasses the limits of our lives. We can imagine things described as belonging to the character, but that we do not possess. If I am myself all the time, I do not consciously feel my body. I seem to know who I am, because of the form I have. When I get sick or experience my body in another way, then I don't really 'feel myself'. I feel somehow like a different person and I will continue to feel like a different person until I get well, and once again take my body for granted. So changing something about the body will immediately give me a different sense of self. This sense of self is what is meant by the use of the word psychology in the Chekhov technique. Changing the body will alter the psychology.

Changing the body sounds like an impossible process, but it is simple, easy and enjoyable. We are always working with the same principles and the same kind of activities; we just find new configurations for them. The principles of Imagination, Energy and Form go into creating the imaginary body.

There are many ways to come up with the image, but only one way to create the imaginary body. It is one more step in creating the vessel for the will force of the archetype to enter into. This vessel helps to refine the power of the archetype into a specific character. It does this by having a tremendous effect on how the movement of the psychological gesture will happen.

The images you choose should serve the character and the play. If you begin to follow the imagination in this way, you can, for example, physically experience the hump that is part of Richard III's body. You will in a sense earn the right to wear the costume, because you will understand the psychology of the man with a hump on his back. It will have a real influence on how you will seduce Lady Anne.

Exercise 18: Changing the body – the imaginary body

From a standing position, bend at the waist and touch your toes. Hang over like that for few seconds. Continue to breathe and relax your muscles. Touching your toes is not critical; if you cannot reach them, it doesn't matter. Relax and breathe while you are bent over. Slowly begin to roll up to a standing position. While doing this, tell yourself that when you come to standing you will be 3 metres tall. You will be a perfectly proportioned person who is very tall. Begin to walk through the space. Sit in a chair. Get out of the chair. Do various things, all the time experiencing yourself as being 3 metres tall. You cannot stretch your physical body into this size, but you can transform the energy of the life-body into that size.

After a while, bend over again and relax. As you slowly roll up again, tell yourself that when you come to standing you will be one metre tall. Do the same kinds of activities, and notice how different your sense of self is compared to a few moments ago, how different it is from your normal sense of self.

It is possible to change any part of the body into any size or shape. It is always a forming of the life-body. Keeping that in mind, change your neck into the neck of a bull. Now move your head with this new neck. Even by changing one part alone, you already begin to experience a new psychology. Change the neck again into the neck of a baby. Play around with different images of your own. See someone who has a body that in no way resembles your own. Imagine that your body is that body. You can try to change the whole body, or you can focus on details. For example, focus on the hands and change your hands. As soon as you change the hands, you must use them as hands, so you can *feel* that you have new hands. Imagine that your hands are made of fine crystal glass. They are glass hands, but you must move them and use them as hands. Look through your pockets or button your shirt with your new hands.

If you can see the hands of the character you will play, and also the neck and the lips, you will go far in finding a completely new psychology. Because we have to use our hands to do the things we do, and we have to use our necks to move the head, and you have to use your lips to speak, a truly new person will emerge. It will still be you, but it will be you in a creative state. You will be able to do things and believe things about yourself that fit the character you are playing. You are completely free in this and it is a great deal of fun to take on new and different forms.

18 CHARACTERIZATION: PERSONAL ATMOSPHERE

If you think about an old friend and try to describe that person in a few words, the chances are good that your memory and description of this person will be about their personal atmosphere. Atmosphere refers to the space that surrounds. In the case of Personal Atmosphere, it means the space surrounding the person. It can be described in any number of ways. Certain ways of describing people make it clear how they deal with the world. The age-old picture of the person whose glass is always half filled, or the other whose glass is half empty is another way of describing a personal atmosphere. We can imagine it as kind of bubble that surrounds the person, and inside the bubble we can put anything we like. We could fill the bubble with laughter for example, so that the space within this bubble, surrounding the person, is laughing. This does not necessarily cause the person to laugh all the time, but this person is more likely to find things funny than not. For another person, the space could be filled with tears. The actor engaged with an image like this is not looking for a way to cry, but is using a poetic and imaginative approach to the character, an approach to a pointed sadness. This bubble, filled with laughter or tears, acts like a filter between the character and the world. What comes to the character from outside comes through the filter, and what the character gives to the world goes out through the filter. It is another means towards consistency; it holds together so many things about the character. It is a great tool for activating the fourth guiding principle.

Exercise 19: Personal atmosphere and the four tastes

We often describe people as being bitter, or sweet, or sour. We are not saying that they taste this way, but we are saying something very specific and clear about them. It is a shared understanding, a way of perceiving those different people. It is an agreement we have and the interesting thing about it is that these people will habitually be this way. This is so because of their atmosphere, their bubble, their filter.

Imagine that the space directly in front of you is filled with sweetness. When you can accept this imagination, then take a step into that space and feel the sweetness on you face and on your chest. Just accept it, you do not have to do anything more. Turn to the right and imagine that sweetness is coming to you. Lift up your right arm and hand to welcome it into your bubble. What is the gesture that would greet sweetness? It has its own quality. Feel the sweetness on your right hand and arm and shoulder. Now look to the left and welcome this sweetness into your bubble. What is the quality of the gesture that would welcome sweetness? Without looking, just know that the sweetness is poised above your head. It is waiting there vibrating, and then it begins to fall on your head and shoulders like a sweet sugar rain. Feel this sweetness landing on your head and shoulders. Then it comes in from behind and touches you on the back of the neck, on your buttocks, and calves. Now you are completely surrounded by sweetness. The world will come to you through the sweetness, and you will act on the world through the sweetness. Allow this personal atmosphere to play you. Do not feel obliged to push anything, or even to act sweet. Just let the sweetness be the filter between you and the world. After a while, when you feel connected to the personal atmosphere, put your attention on the tip of your tongue. You do not need to try and taste anything sweet; this is not the point. The very tip of the tongue is the part of the tongue activated by sweet. This is true for all of us. To imagine a personal atmosphere of sweetness around us lures it to us, and then putting the attention on the tip of our tongue hooks it, so to speak. Using the four tastes is a transpersonal way to engage the actor in you. It takes you out of yourself, and leads you to a character. It is knowledge that is understood by everyone. You can repeat the above exercise precisely the same with the tastes of Bitter, Sour and Salty. The final part of the exercise with the tongue is the only variable here. Bitter is experienced on the sides of the tongue, sour on the back of the tongue, and salty on the middle. Once the taste is 'hooked', it is an extremely easy change to sustain. It seems to take care of itself, and it has the power to colour everything the character does. It should be noted that just because the character is surrounded by sweetness, does not preclude the character from experiencing or expressing any emotion or feeling, even if it does not seem to fit with sweetness. A person with a sweet personal atmosphere can become glum, or angry, or sad, and still have a sweet filter surrounding him. The same is true of bitter, or sour, or salty. You can imagine how a bitter person would laugh. It is not a limited way of looking at things, it is imaginative and it is empowering.

19 ATMOSPHERE: ENGAGING THE SPACE

In our studio we work with space. We begin by believing the space to be a medium that can hold various ideas or images. We then put these images into the space and allow them to come back to us. We always work on the body, making it sensitive to receive impressions, impulses, sensations, intentions. When the images come back to us, it is the body that receives them. The surrounding space filled with an imagination of something specific, say dust, will touch the body. The body will receive this and react. We are not working to make the audience see the dust and say, 'Oh I see the actors are in a dust-filled room because they are coughing.' No. We are working to make some inner, more nuanced psychological connection. How is the psychology affected by this space? We have tried all kinds of substances and fragrances and qualities, heat and cold, etc. We then followed Chekhov's ideas of atmosphere and started to fill the space with feelings and moods and colours. This was all interesting to experience and interesting to watch. We enjoyed our discoveries, but we soon began to run into difficulties. The atmospheres became seductive to the actors and somehow they were led into playing only the feelings or playing the atmosphere, and then all the fascination would disappear from what they were doing. The improvisations and the scripted material became general and unclear.

It is a fatal mistake to let the actors go on playing an atmosphere, because all real intentions and actions become secondary. Chekhov warns us about this. The reaction to the atmosphere is what we should be interested in.

What we did discover, as a result of all the other training to sensitize the body, was that these atmospheres can be perceived by the actor as the space moving upon the body in a very particular direction. For example, the atmosphere of disaster is downward moving on the body, the air itself seems to be falling heavily upon the shoulders and head. This awareness became a real key to finding a reliable approach to working with atmosphere. Returning to the scene, we took out the name of the feeling, or atmosphere, and replaced it with the imagination that the space could move upon the body. This became freeing and somewhat easier to work with. The actors were no longer seduced into playing the mood of a disaster, but were asked simply to react to the space moving downwards heavily upon them. This is easier than it sounds. Once it gets set up, it will take care of itself. The movement really does seem to be out of our control, so we are left with pure reaction to something outside of us. The responsibility to come up with a disaster disappears, and the actors are free to be in contact with each other and to play the actions and objectives that are required of them, and all of this is done while surrounded by a dynamic and energized space. Movement of the space becomes our reality and not a vague idea of disaster, with all the thoughts that can slip into this notion. Thinking will stop things from happening. Anything that is related to movement will always reach the actor as a force.

Exercise 20: Being played by the atmosphere

Walk forward in a decisive and committed way. Know that you are moving in a forward direction. Tell yourself that you are moving forward as you do this. The statement will help you become completely conscious of the direction. Imagine that as you move, the space around you is also moving, it is moving with you. After a few moments stop walking and begin to imagine that the space continues moving, it is moving by you coming from your back to your front. Concentrate on this and let your body become porous. Now feel the space moving through you. It is an imagination but it is easily done. Allow yourself to react to this force moving through you. Do simple things and let the space moving through you begin to play you. It is as if your body is a wind instrument. When the space moves through you in this direction, there is a particular tone that is played. Once you are able to do this, you will notice that it takes very little effort to keep this imagination alive. Soon it will seem to be happening all by itself, the only effort will be to stop it when you need to. Now you are in a position to react to it. It will influence the things you are doing, and also the way in which you are doing them. It is a way to feel what is happening around you. It is a way to create an intangible effect that surrounds the events of a scene. The focus for you is to stay reactive. Then you are not playing anything but simply living in a space that is affecting you. It will not distract you from what you have to do, it will allow you to do these things but specifically in this space made dynamic by the forward direction.

The same thing can be done with moving backwards. In a large open space, walk backwards and tell yourself that you are moving backwards. Make it a very conscious activity. Imagine the space around you is moving backwards with you. After a few moments, when it is clear to you that you are moving backwards, stop. Imagine the space continues to move backwards. The space is moving from the front to the back; no matter which way you turn, it will always have this relationship to you. You must stay consistent in the imagination. Let your body become porous so that the space can move through your body. It enters you from the front and exits out your back. You will immediately notice that this experience is nothing like the previous one, where the space was entering the body from behind and exiting out the front. Accept the idea that your body is a wind instrument and this is the tone that gets played when the space moves backwards. Feel what this is, and react to it. If images and impulses rise up in you, follow them. Perhaps you will feel as if you are in the middle of some circumstance. Play with all of it, welcome everything that comes to you.

The dynamic directions are forwards and backwards, up and down, expanding and contracting. The key is to become conscious of what it means to move in any one of these directions. The activity is to move the space. The focus is to react to the space moving through the body.

Once you have experienced success with the exercise, you can then begin to play with locating the movement in a particular centre. While the space is moving backwards, for example, let it pass only through the head. This is very specific and has a dynamic all its own. You can also pass the space through the chest or through the pelvis. These variations are possible in any one of the six directions.

20 CONTINUOUS ACTING

Acting happens because we are actors. This is sufficient reason to act. Chekhov said it is a mistake to believe that the day you have a job acting you are an actor, and when that job ends you are no longer an actor until you find another project. This is an unproductive use of time and energy. Believing this will kill your abilities and capacities. It will dull your connection to your creative individuality and your talent. The theatre requires increased life, and it is critical that you continually exercise so that you can develop a sense of increased life. We must develop this ability to act without justification, without any outside cause or reasons. Our actor's nature is the thing that is nourished by the technique. It is possible to be always acting, because we receive such pleasure from it. If we desire to do this and follow through, it will lead us to new impressions, new approaches, new discoveries, and new ways of understanding the roles that will come to us. This ability to act continuously will prepare us to have the necessary confidence to be creative. It is really a matter of playing with the technique whenever it is possible to do so. Acting is essentially our ability to give of ourselves. If we do not develop our capacity to give, then we will not know the generosity we need to work from. To work continuously means to find some way to exercise as an actor. Take some piece of the technique and play with it at times when you might be doing nothing but walking, or sitting, or waiting. If we are engaged with our work, we can never claim to be bored, because we can always excite ourselves and stimulate our consciousness by acting. Even to walk, using the Feeling of Ease is to exercise our ability to act. Doing this takes us out of our normal dull consciousness and offers us a connection to the richness that is generally sleeping within us. By exercising like this, we are now doing something with a consciousness that is creative. We can also do errands while we play with the imaginary centre or the imaginary body. We can become the world around us by concentrating on the things we see, and feeling the 'character' of the objects in our world. The Chekhov technique is a grand tool for creative enrichment, because when we use it correctly then we are being led out of our normal sense of self. We put ourselves in contact with creative powers that will always support us. Everything will change, especially the way we see the material of the role we will work on.

5

APPLICATION

Learning this technique is about one's own investigations, one's own experiences, one's own stumblings and successes. Practice can distract us. It is easy to lose the way, to focus on one thing when the etude is about something else. The work provokes many possibilities and the students need to learn where to focus their attention. The teacher must point them in the right direction in their research. If we know what to look for, it is easier to find it. Once the basics are rooted within, then the students can play and choose what they want to take from the technique.

The following section is presented as a way to use the technique in a practical way for actors, teachers and directors. The style of presentation differs from the preceding sections in that I am working directly with actors, giving instruction and direction. It reads differently and so I give this very brief introduction to alert the reader, because things seem to shift gears here. I speak most of the words, but there are also questions and comments given by the actors in the course of the workshops. The many voices that speak up in the course or a workshop or rehearsal are distilled into two, my voice and that of the student (S).

1 WARMING UP

Let's stand in a circle and play a ball game, so we can warm up our bodies.

Here is a ball, please imagine it is very hot to the touch. Simply toss the ball to the partner on your right – don't make your partner work to catch it, aim for their heart. You must catch the ball and toss it on in the circle. Be clear about your giving. Acting is about giving and receiving; this game is a metaphor for acting. Please be sure that you take the ball into your hand, then toss it to your partner on your right. Now we will add a second ball, just keep it going, in the circle from one to another. It is also hot. You must catch it, and throw it. Now a third ball is going, but in the opposite direction. Please don't allow the balls to drop to the floor. Wake up to what is happening. All three balls are moving quickly.

Let's stop for a moment.

It is getting panicked. People are becoming breathless and tense. Stay in contact with the floor, feel your feet touching the floor, and then you will be here and not in some panicked world. Acting requires presence, and this game also requires presence. You really have to be here or the game falls apart. We wake up the body to acting with this game.

That was better.

Now stop. Whoever has the ball, hold onto it, and receive the heat into your hand – just receive it and allow yourself to respond to the heat. Let your muscles be soft so that you can receive it, nothing good will come from tension. Obviously it is not really hot, but it is an imagined heat, receive the imagined heat. Now start the game again.

So we will start and stop randomly. I will call it, and whoever has the ball just receive the heat into your hand and let your body express this discomfort for a moment. I'll say, 'go', and you carry on with the tossing. Use your bodies easily. Lightly express yourselves. The body is all you have, it is your instrument we are waking up.

2 EXPANSION/CONTRACTION

Using your physical bodies, try to become as big as you can, then as small as you can. Begin the movement from a squatting small curled up ball, then grow up and out using all of your body. You are moving now, so listen to your body. Listen to the information that is coming to you because you are making this movement of expanding. Keep breathing as you execute the gesture. Tell yourself you are expanding. You are growing. Please try to feel that it is so. Listen and let the body talk to you. Now move in the opposite way, to contract and return to the place from where you started your expansion.

Becoming as big as you can or as small as you can isn't about a position at the end, but about the journey of it. The real dynamism of it is in the *expanding* or the *contracting*. It is a gesture and gestures imply movement.

Try to be aware that you are a three-dimensional being, that you have a top, and a bottom, and a front, and a back . . . we tend to forget that we have a backside, and so you keep getting bigger and more open in the front, but you eventually start to contract in your back and get tense there, or in the neck because you are tilting your heads back. If it is about getting big, then get as big as you can get and know that you are finished with the gesture, that physically you cannot do it any longer without getting smaller in some other part of your body.

Watch out for the tension, breathe . . . there is a dynamic tension but not a physical tension. Please try to work with a feeling of ease.

What happened to you?

When we are talking about how you felt, nothing is wrong. I could see that different things happened to each of you.

S: 1 *The smallness feels like a hibernation* . . .

2 *Love is expansion and contraction is sad* . . .

3 *One time I felt sneaky and funny, another I was terrified as I contracted* . . .

4 *I didn't have to apologize for taking up so much space.*

5 *When I was open and my chest was exposed I felt unsafe. While contracted I felt safer.*

The different information for each of you comes from the differences in your gestures or the quality in which you did them.

Many of you are thinking too much. Check in with your bodies.

Try to expand and contract like the air blowing up a balloon. Make it a consistent flow, a spherical growing. Arrive at the end of your gesture of expansion, then leave the life-body there, and physically walk away from it.

We developed new eyes in our shoulder blades in a previous class. Do you remember? Just imagine that you have these eyes in your shoulder blades and you can 'see' with them. Use these new eyes to stay connected to what you are leaving in the space. With your new eyes in the shoulder blades, look at the energetic gesture you left behind as you walked out of it.

Repeat that sequence three times. Stay connected to it, own it.

Chekhov said, 'Repetition is the growing power.'. . . what you get from each repetition will grow.

Walk around the room, do different things with this expanded energy that is in you right now. Say yes to even the littlest thing that happens to you today.

More will come from the yes. More will come as you notice what it is, and how simple and powerful it is. More will come when you become aware of the connection between being and moving. This is a *psychological gesture*, the first one we will investigate. The psychological gesture is dynamic because it is moving and therefore useful to you as an actor.

Now do you remember the 'artistic frame'? It is a very special way to move and explore the psychological gesture. The artistic frame means that our movements have three parts to them. The artistic frame is a learning aid; we use it in class, not in performance. It fosters for us a new awareness of moving, because it requires us to move consciously and completely:

1 We begin the movement with the life-body.
2 Then the physical body joins the movement until it can no longer do it physically, that is when we have reached the end of the gesture.
3 Sustain the gesture out into the space using the life-body; this is called radiating the movement.

As the life-body is radiating the expanding gesture, now is the time to walk away from it so that you are leaving something there that is still moving; you are not leaving a frozen sculpture or statue, but a dynamic gesture and you look back at it with the new eyes in your shoulder blades. Own it. Then you can take something from it. Accumulate *the energy of expanding*, don't leave empty-handed. Take it. Make it yours.

When we use the artistic frame, we train the life-body to know how to make the gesture. Then later you will be able to experience this gesture as an inner movement. You will be able to make it without the physical body.

As you walk away from it, say, 'I am'. Speak the truth of this moment.

Really use the artistic frame. Be aware of the three parts.

Now see if you can do this gesture of expanding only with the life-body. Let it happen and see what comes of it.

How do you feel?

S: 1 *Excited.*

2 *Active.*

3 *Happy.*

4 *Generous.*

Now that you can make this gesture as an inner gesture, try to do something else with the physical body.

Do some outer thing as you do the inner gesture.

Shake someone's hand as an outer activity, and expand as an inner activity at the same time.

What do you discover here?

With the right concentration, all of theses things are easy to do. It is much easier to do it than to talk about it. And it is not really worth talking about before you have the experience of it. It is all about having an experience. After that happens, then we can talk about it.

This is an inner event, the psychological gesture. This inner event is translated into an outer expression, the hand shaking. To shake someone's hand in this way is very specific, and it gives something to your partner and it says something to the audience. Something is happening to your spirit and this is reflected in the way you are shaking hands. Work on it in rehearsal. The body remembers everything; in performance we don't have to 'do' the exercise.

Energy as form can be placed in the space. This is one way to work with the psychological gesture. A very particular space on stage can be charged with a gesture so it will resonate. It is something known by your body, because you made the gesture. Knowing it has been put energetically into the space, you can receive its force. You can leave it precisely where you need it to be in the scene. It vibrates there. In an imaginative way it is still moving and is therefore full of its power. You can walk through it and re-experience it when you need it.

S: *What if you keep the gestures with you? The thing behind me kept growing . . . it was bigger and bigger. I don't like to carry too much baggage . . .*

It is fine to have the gesture follow you around. It is what we want, but you can also let it go, forget it if it becomes a burden. I think you will find that it is an ally. Try it.

Now let's work on a *polarity*. You left three expanding gestures in the space; you know where you left them. Collect all three of them. Step into each one physically expanded and transform them. Contract each one of them. Use the artistic frame so that you can know this contracting gesture inside and out.

Look back with your new eyes as you leave these contractions in the space. Say 'yes' to what has changed in you:

- Accept what has happened.
- Play with it.
- Live in it.
- Express it.

Now make the contracting gesture only with the life-body.

Shake hands and do the gesture *inwardly*. Really be in the encounter and really make the gesture. It is possible to do both at once. This is what Chekhov acting technique is: to be outwardly involved in some activity, such as the play, and at the same time to be inwardly alive, because of an image or a gesture.

Okay, return to the first expanding gesture as your inner gesture. Make it inwardly and follow the impulses that come to you.

Live in the now, not a memory of how it felt a few moments before.

What is it now? Let it be what it is, just enter into it and enjoy the freedom.

Begin to go back and forth between these two gestures as inner gestures.

Play with them.

Follow the impulses that arise within you. Try not to second-guess it, just simply follow what is going on in you.

Listen to your body, try to forget your rational thinking, it will only serve to distract you and to fill you with doubts.

S: *I was reacting to the person I shook hands with. If he was expanding I found it very difficult to contract.*

You are a nice and sympathetic person, so you naturally follow the signals that are given to you, this is a fine thing; it shows that you have a sensitivity and you are connected to your partner.

But sometimes the character has to hate the person shaking his hand, and so he could inwardly contract during this encounter. Or sometimes we may hate the other and at the same time lie about it, because it is necessary to disguise these feelings in the play. We have to know what we are going after and then we can find it easily.

Having a technique means that we can do certain required things. If we cannot do them consistently, then we do not belong in this profession. Doing them well and consistently is a cause for great pleasure. Acting for actors always brings us pleasure.

Please try to understand in a very real way that it is possible to say both yes and no with expansion, yes and no is also possible with contraction. This is a very important point that needs investigation. If you miss this, the real dynamism of these movements dissipates and the outcome becomes flat and static.

Love is an expansion and so is rage. It is simply a qualitative difference.

3 IMAGINARY CENTRE

All of us have a centre. All of our movement impulses begin from a centre. The centre is located in the body; it could be any part of the body. As actors, we can choose any part of the body to be the centre.

Let's begin with this image.

Imagine the sun is in the centre of your chest. You can feel the sun radiating up from the chest into the head and down into the legs. The sun is an ideal wonderful image.

Scratch your nose, but let the sun in your chest make the movement to scratch your nose. The impulse comes from the sun. It goes down your arm into the hand. The sun in the chest is moving you; the sun is scratching your nose.

The impulse from the sun now travels down into your feet. This allows you to walk. Find various tasks to do. Let the sun do all of the things you choose to do. Play.

Enjoy the freedom that comes from relinquishing responsibility from the simple things you do. The sun does it all.

Place the sun here, in the head.

Can you feel a shift in the energy? Wait for it.

Scratch your nose from the sun in your head. Let the sun do it. You have more important things to do than to worry about scratching your nose.

Play again. Discover the differences from shifting the location of the centre.

Place the sun in the pelvis.

Scratch your nose, but allow the sun in the pelvis to do it.

Play again and discover the differences.

Let's try the same exploration with ice – the polar opposite of our original image.

It is as if this character was born with a block of ice in the pelvis, or the chest, or the head. Give yourself over to the image. Accept whatever comes to you. Listen to it. It is a psychological change we seek. It is useless to act out that it is actually cold. We are not looking for that.

When we experience ourselves as transformed, then we have made the intended connection to the image. We change the psychology with our imaginations. This may sound like heresy, but Chekhov said: 'We know we are working in the right way when the things we are focusing on take us out of ourselves.' The imagination leads us out of ourselves, it removes our limited conceptions of ourselves and what we are capable of. It increases our possibility, and our power to express ourselves.

Spy back. Ask yourself 'what kind of person was that?'

S: *With the ice in my chest I felt cut off from everyone. I mean it was a character that was cut off. I don't mean I could not make contact, I mean I did not want to make contact. I didn't need anyone else.*

You have to move to find the behaviour. Behaviour does not come from sitting around feeling things. This kind of thing leads to paralysis, where you become stuck in the feeling world. How will you interact if you have this sense about yourself that comes from the ice in your chest? It is interesting, you have now created for yourself some kind of inner obstacle. Perhaps you are in conflict, yet you have to deal with the world.

Move in a new tempo – strong staccato. You will be given everything you need. Justify the tempo shift. Stay connected to the centre. You will know the cause for your shifting the tempo.

Change the centre. Put the ice in your head. Make contact with it. Place it there. Feel the shift.

Do simple gestures that lead to more complicated ones.

If you give yourself over to it, it will tell you what to do.

Allow it to free you up.

Is it a different person when you change the centre? What is your experience? You are changing, and playing with the centre. Let this be the centre of your being, the centre of the character. Everything you do comes from here.

S: *Can you connect sun to expansion and ice to contraction?*

Of course, that is the reason we work with those particular images and that is why we introduce expansion and contraction early on; it is a principle and a lot of things are built upon it. The sun radiates outwards. It is giving and nurturing. By its nature it is expansive. Ice by its nature is psychologically contracting.

These things are tied together in a really beautiful and simple way.

Everything is tied together. The technique offers multi-functional and simple tools applied to dynamic principles.

The centre now is razor blade eyes.

The eyes are not being cut; the eyes are razor blades. All movements are coming from there. The razor blades are doing it for you.

S: 1 *This feels like a sharp and cold person.*
2 *Calculating.*
3 *Aggressive.*
4 *Isolated.*
5 *Observant.*

The image is narrow and contracted.

Keep the same location, but change the image to candle flames in your eyes,

S: *I feel loving, soft.*

The flames are soft and expansive. They embrace everything.

S: *What is the method of crafting? Say the character is very gentle and you want to use eyes like candlelight?*

This is fine. This is the centre of the character – you would have these eyes the whole play. You couldn't feel the character without this. It takes courage to do this. You have to say, 'tonight's rehearsal for me is about finding the centre', and you have to give it a lot of attention for a few rehearsals, then it finds its place and you don't have to concentrate on it any more. It will be there.

You have to work with your intelligence, but not your thinking. You can't say the character has flames in the eyes, because it felt good in class. Your imagination will give you the image. But for now just use this one and find out what it means to you today.

S: *Can the image change?*

Yes, you will know by playing with different images in rehearsal. It is possible to change the quality of your centre, just concentrate and everything will come from that place. The character will show you the quality of the image: brittle, soft, sharp, dry, moist, etc.

S: *Can the imaginary centre change?*

If you keep moving it all over the place, it is no longer a centre. But yes, you can do anything you wish to do, anything that will help you towards creating your performance.

If you think too much about these things, it leads to trouble. The rational part of the brain, what Chekhov calls the 'little intellect,' says, 'this is not possible, I won't allow these things to happen'. If you try to reason it out, you kill it. You must see the character in your imagination moving around in a particular way. You can determine, 'Hey, it appears as if he has a broken glass bottle in his pelvis.' It is not about you expressing the pain of it, it is about you being born with a broken bottle in your pelvis. This is your life. At the very least you must try it, because your imagination gave you the image, you must trust it and give yourself over to it. You can always discard an image if it does not feel right. Working with the imagination is free, and if a particular image does not serve you, it is very easy to replace it with another. On the other hand, if you work only with your reason and your thinking, then it is very difficult to discard your findings, because your reason will find many ways it can justify why it should serve you. Directing actors who become fixated on ideas that are not serving the production take up a lot of time and waste energy. Really, it is a lot of work to act being led by your thinking. Thinking, as wonderful a process as it is, is kind of narrow as an activity. Thinking needs to narrow things down. Imagination is broad and expansive as an activity, it wants to include many things.

The most difficult thing about acting is to know what is necessary.

We learn techniques to arrive at the necessary things, but we as actors need to know what is necessary for the scene, the play, the character. You can express anything you want as an actor, but you must be able to know what it is you want to express.

The script says what is necessary. At this particular point in the play, we must cry, then laugh there, etc. That is what is necessary. We must arrive at the things and come clean in between. It is best to stay simple in your ideas. Proper use of the technique will complicate the rhythm, leading you to complex compositions.

4 THINKING, FEELING, WILLING

Let's play a different ball game tonight. Everyone has a ball, so let's toss the balls all at the same time. Each of us will have a person to throw the ball to and we will always throw it to that person and we will always catch it from one. The ball I catch will come from one person and the ball I throw will always be thrown to one person. It is a triangular connection. The responsibility is on the thrower. Aim for the heart so that the catcher can always expect the ball to arrive in front of his heart. Each of you is a thrower and so each of you has the responsibility.

Try to keep all the balls in play all the time. All balls that fall to the floor must be picked up and reintegrated into the game. Keep all of this together as the game continues. We will throw and catch continuously. Let's see how long we can keep it going.

That was good, first a bit of a struggle, then we found it and could sustain it.

Do you feel more awake to the group and to your body? Good.

Have you noticed any sticks, balls, and veils out in the world? Once you notice these things in your daily life, the method is accessible and practical for you. This business of stick, ball and veil is quite real. You will see it if you look for it. Have you found it in yourself? Try to become self aware of this. You should be able to distinguish between your own qualities and the qualities of your character.

S: *I think I might be a feeler first. Once I thought about going through the world . . . (gesture to the head).*

Not a thinker? It's okay to be a thinker. It just is. It doesn't mean that you do not feel. Just as a person who thinks, as the last function, like Romeo, is not a stupid person. It just is.

It's crucial to know what type you yourself are. For some parts of the character, you can go directly from yourself, for other things you have to use the technique.

We are talking about your life. What you are in your life. Not in your work.

Let's look at something you already know, to see if you can discover something new in it. Behave as if you don't know this information, even though it is obvious that you do know it.

Please feel that you have a head. Try to feel that the head is a form. It is round, and it is nothing like any other part of your body. This round head is where you do your thinking, but for now just try to feel that you have a head. Connected to this round head is a vertical line that we call a neck. Please try to feel that you have a neck, let the neck feel the neck. With the neck you can move your head. Please move your head using your neck. It is possible to say something by moving your head, something that you cannot say by moving your hands or feet.

Okay? Now move your head and say something very specific with this moving head. Do it again and again so that it is clear what you are saying. Now do this same movement very quickly, I mean really quickly, a strong staccato tempo. Quick, then stop.

Are you saying the same thing you said a moment ago? Really try to do this with a feeling of ease. As you make the movement more quickly, notice how it becomes a bit more difficult to do it. Tell yourself to make this quick movement with a feeling of ease. Try to experience doing this movement with a conscious feeling of ease. Now do it very slowly. Legato, slow with no stops. Notice that it becomes easier to make the movement. Are you saying the same thing?

Connected to the neck is a horizontal line, we call this line our shoulders. Please try to feel that you have shoulders and that this part of you is a horizontal line. There are things you can say with your shoulders that you cannot say with your head. Move your shoulders and say some thing very specific with the movement. Now do the same movement quickly. Again tell yourself that you wish to make this movement with a feeling of ease . . . Try to take pleasure in the fact that you can move so quickly yet so easily. What are you saying with this movement?

Now make it slowly, very slowly and see if what you were saying has changed.

Connected to the shoulders are two vertical lines we call them arms. Try to feel these arms, you can move them and while moving them it is possible to say something very specific, something you could not say with your legs. Repeat the movement and be clear about what you are saying with your arms. With a feeling of ease, make this movement very quickly . . . What changed for you there? Now make it slowly. Let your body talk to you, listen to what is available to you now.

Connected to these vertical lines are what we call hands, look at your hands and admire them, then look at your hands in your imagination and try to feel that you have two of these great things we call hands. You can move your hands and say something very specific, something you could not say with your knees. Make a simple movement with your hands that says something. Then change the tempo from normal to staccato, then to legato . . . Play. Listen to what the body is giving you as knowledge . . . Feel the wonder of having hands and using them.

Also connected to the shoulders and the neck is a great mass coming down like a cylinder, we call this the torso. In there are our vital organs, especially the heart protected by the ribcage.

Please try to feel this part of you. You can move this mass and you can say something specific when you do. You can articulate it and make little movements, or you can make very large movements that require commitment. Make a movement that speaks and says something very specific. Change the tempo, realize that you have a body and that your torso is now speaking.

Connected to the torso is another horizontal line. This line we call the pelvis. Please feel that you have a pelvis, put all your attention there and you will know right away what kind of power lives there. Feel this part of your body as it separates from the torso and can move in a completely different way. You can move this part all by itself, and you can say something with your pelvis. Say a simple thing by moving your pelvis. Make it simple and clear, it will be a beautiful thing because it is so simple . . . Wonderful . . . Now change the tempo and play as before. Feel how it is to have a pelvis.

Connected to the pelvis are two long vertical lines, we call them our legs. Feel that you have these strong and nimble elements in your body. Feel how they transport you and hold you up, feel how easy it is to move them. Use your legs and say something. Repeat it. Then using a feeling of ease, make the movement quickly, and so forth. Play with the movement . . . Discover something new about having legs.

Connected to the legs are these miracles we call feet. Feel that you have feet and how they connect you to the earth, feel the earth contacting your feet, feel them as a part of your body situated so far from your head and neck. Walk using your feet and notice how you push off the earth as you walk, that gravity causes you to fall, and then you catch yourself with your feet giving you forward momentum. You can stop walking and use your feet, or one foot, to say something, a thing you could not say with your head. Play with the two tempos, change them and discover something new about your feet.

Now walk around the room and feel yourself as a whole being, with this human form possessing all these parts. Feel these parts functioning together as a unit when you move. Make a large gesture using all of the body. All the parts move as one integrated whole.

Good . . . The gesture is full with the body. Try it now using a feeling of ease so that the gesture is free, and large, and easy . . . Radiate this gesture. Use your entire being to make this gesture. Use different tempos now. Feel that your movements are harmonized.

Surprise yourself. Listen to your body.

What does it mean to do this? Make the gesture and as it is radiating say this short piece of text, 'I have a body, and my body is expressive.'

Say it again even though you think you already know it. The words will affirm the feeling coming from the gesture, whatever it is.

Okay, lets talk for a bit.

Look at what just happened when I said that. Did you notice the shift in our postures? Did you see how we each slipped into some sort of 'listening pose'. Did you notice this habitual way of standing?

No? We don't generally notice. We unconsciously slip into these poses of habit, in effect to deny that we have bodies. Generally these poses are quite contracted, and often tense, but we don't even notice how tense they are. In this tension we leak energy. We immediately leave the sense of having a body and become heads to consider what is being said. Just a moment ago you were very much in your bodies, present in the space, radiating energy, feeling your self to be an integrated whole. Now we are a bunch of heads.

I am not trying to take that away from you, you can stand any way you like, but habits like these take away your physical consciousness.

Let me suggest to you that while we are in this room together, every five minutes or whenever you remember it, do something like this – put yourself in a pose that is silly or ridiculous, soft and easy, never tense. It is a fun little game we will all play together tonight. Let this be a reminder to yourself that you have a body. And when you are in this soft and silly pose, say out loud, 'I have a body.' It is okay with me that you do this in the middle of any exercise, or afterwards or when I am speaking, or even during an improvisation. We will all understand, and we will encourage each other in a new physical awareness. Try not to repeat the same pose, because it will just become a habit again. Let's try to stay with it and change it each time. We are working with character, which requires a real desire to transform our sense of our bodies.

Here is some information you may find interesting. It is not specifically Michael Chekhov, but it is helpful in finding the imaginary centre:

– The top of the head is the thinking part of the thinking centre.
– The jaw is the willing part of the thinking centre.
– The eyes – feeling part of thinking centre.
– Hand – thinking.
– Fingers – thinking.
– Centre of hand – feeling.
– Heel of hand – willing.
– Forearm – feeling.
– Bicep – willing.
– Elbow – willing.

5 *DESIRE UNDER THE ELMS*

Now these two characters, Eben and Abbie, what types do you think they are? There are three characters in this drama. What if we assigned a dominant function to each of them? Even though no one will be working on Cabot, it is not a bad thing to consider where he fits into the *whole* along with the others. You have to be able to look at the whole. Once you have a *sense of the whole*, you can then look at the specifics, tear it apart if you need to. But you must know the whole in order to put it back together. It is like when you do a jigsaw puzzle, you keep looking at the box to see the picture, the whole picture.

S: *I think that Cabot must be a willing type because of everything he has accomplished in his life, the farm and the money. It was all his doing and he makes that point to everybody.*

That seems simple and clear enough. I would agree with you, he is a willing type. Already we know something about this character and how he functions in the whole. So what about Eben? What type is he?

S: *He is so stubborn with both Abbie and his father, I would think he is a willing type. Can we have two willing types in the same play?*

Yes, you can have as many willing types as is necessary to tell the story. But where do you think this stubbornness is coming from? What is provoking it in him? We can see it from the very beginning of the play.

S: *He must be a feeling type. He feels cheated by Abbie's presence, and he hates the old man very much. He loved his mother tremendously. All this we know right away. These seem to me to be strong feelings.*

I agree, I too see him as a feeling type. Remember to be a human being, all three functions need to be operating. We are interested here in finding the order in which the functions do operate. We can see without analysis that Eben operates first from his feelings, then he acts, and finally he thinks. This is huge, this information properly applied can open many doors for the actor playing Eben. It can solve problems and be a source for real creativity.

So, what about Abbie, what type is she?

S: *Would it naturally follow that she is a thinking type, just because the other two are willing and feeling? Do we have to make her a thinking type by default?*

No, you choose what works best in telling the story.

But let me say this: Eugene O'Neill is unquestionably a great playwright. It often does follow with such writers that we can find this neat little arrangement in their plays. Whether it is conscious work or not for them doesn't matter, this kind of revelation often appears in their work. It is the interplay of these three functions that make up life or drama. So it is helpful to look for it, and to recognize it when it is there. This business of thinking, feeling and willing creates marvellous containers to hold the work in.

What do you say about Abbie? Is she a thinker, a feeler or a willer?

S: *She comes and she wants the farm, and everything seems to be about getting the farm any way that she can. She marries Cabot, she moves in, she takes over the wife and mother duties, she works on Eben. I see her as a willing type.*

When this character is revealed to us, she tells us a story about her past, her dreams, her thoughts about other people, and about herself. We come to understand that she has a plan to take this house for herself. She thought about marrying Cabot; it was not a spontaneous act. You can see her calculating all the time how she can get through to Eben. She understands him because she thinks about him. She tells him he is going to succumb to her long before he does, because she envisions it happening in her plan. Can't you see her as a thinking type? . . . Yes?

So here it is this arrangement of these three characters in this drama. This will help us on our physical journey towards realizing these characters.

Now use the stick, the ball and the veil to explore these characters. Here we will use a stick for Abbie and a veil for Eben. What type of veil could he be?

Remember the movement of the veil? It is soft and fluid like a water plant. You have so many possibilities for an image to work with. We know from reading that he is socially awkward, and he is a farm boy, his father complains how soft he is compared to himself, yet he is durable enough to work this very difficult piece of land. Take all the facts that have come to you by reading the play and try to make use of them to find this particular image of a veil that belongs to Eben.

The women should do the same for Abbie. Find the stick that can physically capture her. The movement is linear and a bit rigid or clipped. This is very preliminary work on the character, it is not definitively the character, it is just the type . . . The specific stick or veil will help you get closer.

Concentrate on the image and then you can incorporate it. You must put it into your body.

The steps of the process to creating as Michael Chekhov sees it are:

Imagination – Concentration – Incorporation – Radiation – Inspiration.

Here we are at the beginning of the work and already you are warm with your questions and answers. You are moving and trying to find with your bodies these simple things and this makes you active and accepting. The work comes to meet you and you are engaged in physical investigations.

The psychology and the body are intertwined. By moving in this way you will discover and acquire things.

This warmth in the work is prompting your desire to act, because it comes from your creative individuality. When you incorporate the image you are really living with it in your body.

You want to be able to say to yourself 'the image has me'. You have to give yourself to the image, really move your life-body towards the image. Then it will free you, and surprise you, and inform you.

6 ACTION, PSYCHOLOGICAL GESTURE

Say this sentence: 'Drink the water, it's good'. Say it many times so that you know the real meaning of the words you are saying.

Now make a large gesture that expresses the essential intention of the words. One movement, not a pantomime. Try to find the essence of the idea, the feeling of the whole of it. It is very simple, so it should be easy.

Ask yourself if the movement is really expressing the whole of it. If not, you have to change the gesture.

Use the artistic frame. Let it start with the life-body, then the physical body picks up the movement, let the life-body continue the gesture beyond the physical body, and as it is radiating, speak the lines, 'drink the water it's good'.

S: *I feel energized when I do this. Like I really am connected to something.*

We have to have streams of energy behind this work. The gesture opens up these streams.

S: *I don't know if my gesture is too literal or not.*

Do you feel empowered by the gesture?

S: *Yes.*

Then, it is good for now. You will find a better one later.

Now make the gesture only with the life-body, keep making this gesture with the life-body, make it, sustain it, then speak when you absolutely have to.

You have to let the life-body prepare the gesture and do it first before the physical body. You are training the life-body to know the gesture. We are using the physical body to find it and the life-body to learn it. After we do this, then the life-body can do it and use it.

What are you doing when you make this gesture? What do you feel like you are doing?

S: 1 Offering.
   2 Embracing.
   3 Giving.
   4 Demonstrating.
   5 I am saying 'Trust me'.

What are you doing when you ask someone to trust you?

S: I am giving them a piece of myself.

See, what we are doing here is finding the archetype. We are going about it in a different way – through the back door. This is not the archetype for the character, but for the action. Actors often call it the objective.

This is one of the archetypal statements of action. It is Giving. 'I give' is the statement.

You can give someone a punch in the face and you can give someone a kiss on the lips. The simple line 'I give' is the line of action. This is one of the six archetypal actions.

The gesture is one movement, make it as active as possible. Use as much of your body as possible. Do you feel streams of giving going on inside of you?

S: I feel something, but I cannot identify what it is.

Then follow the impulses that come from this movement.

When you are working the gesture with the physical body, allow the life-body to radiate before you speak.

The psychological gesture is a kind of question, and the answer comes in the radiation – when the life-body radiates the gesture. We are making an energetic wave for the text to surf on.

When you make the gesture as an inner gesture, with the life-body alone, then you can speak the lines at the same time you are making the gesture, because you are involved with the energy and you are radiating that energetic wave.

You can name any action, and you can find it in one of these six actions.

All the actions are connected in some way: I give, I take, I want, I reject, I yield, I hold my ground.

What are you doing when you do this – this giving? If we want to work with action, we want to use things that are tasty to us.

S: *I felt like I was shitting on her, 'I shit on you'.*

That is a form of giving.

So you have to start by identifying what you are doing to the other person – shitting on her, then you see what shitting on a person is – it will be one of the six archetypes of actions. In this case it is a giving for sure. Do you see that?

If you go straight to the archetype, then it becomes heady, intellectual, dry and general. Find the precise action and it will lead you to the archetype and to the correct gesture.

Let's just look at this for a moment and see if we can find the archetype of a specific gesture. Name an action:

- To spy = I take.
- To flirt = I give or also I take.
- To challenge = I give.
- To kill = I give or I take – depends on what is going on.
- To redeem = to lift/I give.
- To argue = I hold my ground.
- To comfort = I give.
- To plead = I want.

Chekhov gave us five archetypal gestures to investigate this world of action:

To push — to pull — to lift — to throw — to tear.

It's the quality of the gesture that gives the specificity of the action.

Live in the gesture and follow its impulses — sustain the gesture for a little while and then outwardly come to life by following the impulses from the inner movement.

S: *Each response is so specifically different — pulling from the pelvis, as opposed to pulling from the head — Is it a formula?*

It's a palate. It's a means of finding what is necessary. Knowing what is necessary, that is what the difficult part is — when you have found that, then you have the tools to get it.

S: *I have to boast and I don't know what to do.*

How does one boast?

S: *Should I find the action first or the gesture first?*

They are the same thing. In the end, it has to become an inner gesture. You must do the gesture inwardly.

S: *So is that the centre too — you can put it in the centre as well?*

Yes you can. Why not try it?
What are you boasting about?

S: *Coming from little and now owning lots of stuff — winning.*

Boasting is also giving. Can you see that?
Expand into that gesture. Expand into 'I give'.
You have to look at everything as movement. For instance, what is the movement of love? It is giving and expanding.

It is always movement towards the beloved. I am continually moving towards my beloved. Sending out of the heart, so to speak. Each lover sends their heart out to the other and they meet in the middle, they actually find each other outside of themselves.

If it is really love, then you trust that it comes back. It is all movement.

The earth is in movement, air, water; it follows certain laws. As creatures of the earth, we are subject to the laws of movement.

S: *I think that she is seducing him in that long speech about the heart.*

Seducing. What is the gesture of that?

S: *Pulling? Taking!*

Yes, it is a constant pulling in a secretive, sensual and playful way.

Pushing and pulling are very useful and basic things.

We immediately feel yes and no in them. It is possible to improvise the whole scene with 'yes' and 'no'.

Here is the working progression of the gesture:

1 Find it with the physical body. Develop it so that you can feel that it excites something in you. It should excite the very thing you are looking for.
2 Now use the life-body to start the gesture and follow with the physical body. When you have physically reached the end of the gesture, then sustain it by radiating the gesture with the life-body. This is the artistic frame.
3 After a while, try to make the gesture only with the life-body. Follow the impulses or the streams of specific energy that are moving in you. Your body will be alive with giving or taking or whatever it is you are doing.
4 Find the quality. It will be the quality of the movement that will give you your specific action.

Use the gesture of 'I Take' that you have developed, now make it carefully. When we work with quality we have to do it 100 per cent. One hundred per cent of the effort is put into doing the gesture *carefully*. This is very engaging, see where it leads you. You know the gesture so you can concentrate on the quality or the *how* of making it. Now this gesture is very specific and it wakes up in you a specific energy that is easy to follow, so follow it, say yes to it.

S: *Doing the gesture carefully made me very aware of my surroundings and that maybe I was not so safe.*

Are you actually talking about you or the character not feeling safe? I mean, was it pleasurable or not as an experience?

S: *Yeah, it was really pleasurable and inviting. I could work with that.*

Then that sounds like a good thing, an artistic thing. That is what we are after. We want to find the pleasure in acting to make it a very creative event.

Practise this sequence:

—   The gesture of 'I take'.
—   Careful taking gesture.
—   Careful taking gesture from head (thinking centre).
—   Inner gesture – make one little outer movement that comes from the inner movement.
—   Make the same gesture from chest, (feeling centre).
—   Make the same gesture from the pelvis, (willing centre).
—   Now change the quality to carelessly.
—   Move the gesture carelessly from the thinking centre.

Now make it an *inner gesture*. What is the impulse? Follow that, it is yours to pick up and use. Do it 100 per cent carelessly.

In any action there is a moment that could be called the high point.

We could call it the sweet spot.

Practise the gesture now with your full attention. Where specifically do you experience the high point of taking from this gesture? At which moment do you get the most energy from it, the biggest kick? It can be in a different spot for each of you. Pay attention and try to find it.

When we work with the life-body, we are able to enter into a fantasy time/space. We can sustain a very little movement for a long time.

If you look at the whole gesture, it has a beginning, middle and an end. Because it ends, it is difficult to sustain. You feel compelled to repeat it over and over. There is not so much there to really excite you continuously. When you find the sweet spot and live in it, there you will find vitality and creativity.

It is possible to live in a small amount of movement for ten minutes. Allow yourself to do something outwardly that comes from the inner gesture. As you practise, try to avoid speaking until you feel compelled to.

Follow the impulses so that the inner event is translated into the outer expression. Don't hold the high point. Let it be a continuous movement, but within a small range. It is still an event. It is imaginary and that is why we can sustain it. It is not replaying it again and again; it is living in it.

Let's try this simple exercise to demonstrate what I mean. With both of your hands make a fist and squeeze these fists harder and harder. Notice what kind of change comes over you because you are doing this.

If you continue to squeeze, then you will become tense and everything will die. This is interesting and it should tell you something about staying free of tension. Energetically the body will not work for you if you are tense. You will not receive the impulses, because there is no way for them to move through tense muscles. Now using your inner hands, make fists and squeeze them harder and harder. The outer hands are free of tension. They are not fists. Now you are sustaining a little movement, an inner movement. Continue this for a few minutes. Say yes to what you are experiencing as a result of this inner movement. So long as you are engaged in the movement and free of muscular tension, you will receive information, and impulses, and sensations.

If you analyse the making of a fist, you will see that it is a very small movement. Inwardly you can be living in the moment continually. Do you agree that it is a continual experience? Do you see how easy it is to do? This is not a complicated thing and it gives you so much.

S: *I feel like I want to explode, like I am a dangerous person, someone who needs to be reckoned with.*

Is your body tense? No? Good. This is how we can work with sustaining the gesture.

Let's return to the gesture of 'I take'. Do this gesture sensuously. Don't get all 'sensuous', then move. Find sensuousness in the *way* you move.

Now find the *sweet spot*. Be clear about this, you will find it if you are looking for it. Once you have found the sweet spot, do it with the life-body so that you are continually living in or experiencing the sweet spot. It is simple and quite possible to sustain. Notice how you have become filled with the desire to take, the need to take, and you know how to take because you are always taking inwardly. Your body is very alive because of it. You have found your way to an increased life. You are expressive and free with your body, not stiff and dull with decreased life.

The body remembers everything, this is why the gestures are worth working on. They will be there for you when you need them. The mind can forget and the mind can lie, but the body remembers and the body always tells the truth.

Every little thing you put into the gesture will affect the message you get. If the head is down or up, if the hands are opened or closed, if the arms are above the heart, at the heart level, or below the heart.

Eben puts a lot of energy into rejecting her.

The archetype of rejection says that I am finished, I turn away from it, I will never look at it again, I am rejecting it.

Some of your gestures are interesting. You all have to explore them. Even though it is interesting, it might not feed you in the right way; you have to listen closely to it. What does it mean to turn your back and head, and lean down with the hands covering the face, with one hand blocking the sight of the other person/thing? How is that different from just standing erect and blocking with one hand and turning the back and head? I am sure you will find a huge difference between these two rejections. Which one is appropriate for him in this scene?

In time his resistance weakens, he continues to resist but a bit more feebly, because he feels as though he has to do it. He is sworn against his enemy. But she begins to break him down with all her different forms of giving, and a few takings as well. Let these streams of rejection filter throughout the body so that something happens. It is nice to have this experience, but find out what it means. Let your body find out what it means.

In the second scene in the parlour, he has lost that battle and a new one begins within him. He has resisted her so much that when he finally falls for her it is with a great force. All his resistance has produced a kind of dam in him, and in this scene the dam breaks.

The *who* is the character, the *what* is the wanting, the *how* is the sensuousness.

Try to see the movement, see it as an abstract, as a dance, if you will; try to understand how these people move together. Who is doing the pushing, when? And who is doing the pulling? What is their dance?

7 QUALITIES OF MOVEMENT

We have to work on quality today, and we will work on it in a very pointed way, actually in an archetypal way. Chekhov gave to us through his book four distinct ways to move and to explore movement. These four qualities are marvellous, because when you work on these four you are working on countless ways to move. It is important to look at this now, because you have to bring quality into the psychological gestures you are making. As I said in the last class, quality gives you the specific and required action.

These four qualities correspond to the Greek elements: *Earth, Water, Fire and Air*.

Let's work through them from densest to lightest. We will start with the earth, this quality of movement Chekhov calls, *Moulding*. Imagine you are standing in wet clay. Feel its resistance as you try to move in it. Chekhov uses the word moulding to describe the movement, but sculpting might be more appropriate, because you will move as if you are sculpting the air itself. The air around you is this clay, and you have to sculpt or mould it. Feel the air resist your movement as you sink to your hips in the mud and then up to your torso and arms and neck. As one movement ends another one begins. Breathe, breathe, breathe. Maintain a feeling of ease as you 'mould' your movements. You are working with the resistance but try not to get tense while doing this. It is possible if you work with this feeling of ease. You can take up space. You can move past the limits of your ability. Now that you are fully in this clay, sculpt huge geometric shapes in the space. Don't be tempted to use only your hands. Use your whole body. You have a neck and you have legs and feet and elbows, etc. Be sure you understand that to move in this way is not just to move slowly, but to move against a resistance. When you feel that your body truly understands this way of moving, then let the quality of movement disappear from the outside, but increase it on the inside. Your life-body is now moulding. Find a chair and sit in it. This chair is not comfortable, try to find comfort by moving in the chair, inwardly mould all the movements, just let the physical body follow your inner moulding. Find how quickly you can continue to move outwardly, yet stay connected to the inner moulding. Keep increasing the outer tempo and stay in contact with this inner moulding. Just sit there for a moment. Look at something, point to it and say, 'Look'. Inwardly mould the movement of pointing. What does this give to the gesture of pointing? Does it take on a particular value for you?

The technique is all about movement. You will be moving all the time. See what psychological qualities you can absorb from the movements.

What does this feel like?

S: It is difficult.

What is difficult? To move your body like this, or to move your life-body like this?

S: How can I use this? I think I have difficulty understanding the value. I don't think I would ever move like this.

You have to remember that we move in order to absorb psychological values from the movement. Anything you discover by moving the body in a particular form, or moving it in a particular manner, can be reawakened by inner movements or qualities. I think what is making it difficult for you to do it is that you are thinking about it before you do it, and so you allow doubts and intellectual resistance to creep into you. Please try it again and put all your attention on how your movement is resisted by the space. Feel the kind of effort required by the body to move against this resistance. We will find a way to put it to use later. Just experience what happens to you, to your body and your psychology.

Turn your head (moulding) and look at someone, and then turn again and look at someone else.

Try speaking now. It is possible to mould the lines. Use a monologue you are familiar with. You will be speaking this monologue in a new way, don't worry if it feels strange, just try to do it and see what comes of it. It may not be the way you want to do it, but then again you may find something new about this text that you had never considered. It is just a playful experiment. Imagine that the words come out of your mouth as formed words . . . they have shape and substance in the space. It is almost as if you can see the words coming out of your mouth as already formed things.

How was that? Speaking like that?

S: Interesting. But as I formed the words I felt I was focused on the words and lost the context.

Could you repeat that? I didn't understand what you said.

S: As I form the words I felt I was focused on the words and lost the context.

I'm sorry; I'm not getting it. What did you say?

S: I said, as I form the words I felt focused on the words and lost the context.

What?

S: I lost the context.

What? I'm not getting it.

S: I said, etc., etc.

I'm just playing with you. Did you see how she started to speak to me? How she began stretching out her words in order to become clear? In a literal sense, what she was saying was being met with resistance, (I didn't understand her) and so she naturally began moulding the words so that they were clear and then I would understand her.

Please don't think so much right now about these things, there will be time for that. Try to find them as simple things and discover what they could mean.

I just led you into a natural way of using this. Do you recognize it? It makes such a simple sense, does it not? And it is quite possible to do it, to really do it and then receive from it wonderful means to express yourself. It is simple but it is not primitive. It is an elegant way to work on form, and to bring some kind of order into your work.

Speak again and mould the lines.

I will clap my hands and you will go back to normal speaking. I clap again and then change back to moulding.

What happened here?

S: *Direct, pointed. Like pounding the other person in the head with the words.*

I form the words so they will be clear.

We have to be able to make use of this. It is always about working as an actor. About the other person being affected by what we say or do. About how specific we need to be.

Make this movement with your head (turning from right to left). Now don't move your head but move the life-body head . . . the 'inner head' meets resistance.

What do you feel is going on for you? Translate the inner event to an outer expression.

S: *The action informs the story. The relationships changed.*

Walk to get a sense of the resistance of moulding, but it doesn't have to be slow. Walk as quickly as you can with the resistance.

When I ask you to mould, it doesn't have to be slow. The inner movement is moulding, the outer movement is quick. What happens then?

This is great for film, just do the inner thing, the camera will pick up all the inner quality.

Let's move on to another quality. We will call it Flowing. It is connected to the element of water. As you sit in your chair, imagine that you are sitting on a large stone in the middle of a river. This stone interrupts the flow of the river, but the river passes by you on either side. Now stick your hand into the flow and feel it take your arm. Then forget it.

Keep sticking your hand into the flow and feel it take you. Eventually you will get up from the chair and be taken up by the flow of the river. Now you are in a flowing river and you cannot stop moving because the flow just takes you. Notice that your movements have no beginning, no middle, no end; experience the movements just flowing on from one to the other . . . Stop. Feel the energy flowing around you . . . Now you sit in the chair but you can't stay still in the chair because it is not comfortable. Try to get comfortable in this chair by moving, you are flowing so you can't stop moving . . . Stand or sit, but something is always moving. Perhaps it is just your thumbs twiddling, scratching, looking through your pockets, etc.

Make the movements as small as possible, but something is moving all the time, there is no way you can be still, move only the eyes, get up, live in this flowing restlessness.

Remember we are here to absorb psychological qualities from the movement, to make use of it a later time.

Now let's return to your monologue. So when you speak, it is like you turn on the waterspout and the words keep coming out of your mouth. You can't stop the flow of the words. It is incredible but possible, they just keep coming out of your mouth in an endless flow. It can be fast or slow, but be clear that it is an uninterrupted flow.

How does this feel?

S: *This is much easier than moulding, but it is also quite different. Things don't seem so important.*

Flowing engenders a feeling of ease.
Moulding engenders a feeling of form.

S: *Characters pop into my mind. It is easier to be easy.*
Not thinking actor-y things . . . but actor-y things are coming to me. I felt like what I was saying was less dire.

Good. Start speaking in a flowing manner, when I clap, change to moulding, clap again, then change to flowing, etc.

S: *This was so interesting; the monologue seemed just naturally, suited to the shifts that came from clapping. How is that possible?*

Well, I have to say that the clapping was random. I don't know your monologue. Perhaps some of it was correct. And maybe not every one of you felt that it was 100 per cent correct. It is just an exercise, a way to play with text and discover something that comes to you as a gift, not always thought out. Thinking so much can really become tiresome.

So let's make our way through the elements, getting lighter and lighter. That means that fire would be the next one. This quality as Chekhov named it is Radiating. This means that light is coming off you, and you are consciously sending it out, continually sending it out.

Imagine that there is an aperture on top of your head, like the iris in a camera and it can open and close, as you will it to. You must begin this exercise with the idea that you are a luminous being, and within you is a bright light, and it is just waiting to come out of you. Open this iris on the top of your head and feel that the light is shining out of this opening. We will start on the top of the head so you won't be tempted to try and see this light, which you cannot actually see, but you can certainly feel it as you are doing this. And this is what we want, to feel it. Know that you are lighting up the ceiling, or if you bend over to tie your shoe that you are lighting up the person in front of you.

It is a wonderful feeling of power. Imagine you are in a dark room and that, because of you and your radiation, it is possible for others to see their way.

Now close the iris down completely. Do you feel the difference between having it opened or closed?

S: *There is a huge difference but I cannot say what it is. It makes me feel quite present and strong and alive when it is open.*

Radiating is a very pleasant activity. The world becomes more attractive, because we lend it our light and this makes things appear bright and pleasing. It really is a kind of power that we can use.

We can put an iris anywhere we like. Put one in the palm of each of your hands, and experiment with opening and closing them, or opening them half way or one quarter. Try to have one hand closed and the other open, etc,

Look at the different variables, don't forget the iris on your head. Try putting them on your feet so that when you walk you are lighting up the floor with each step you take. Now put one on your tail bone. This makes six irises that you can play with.

Open all of them now at once, and then close them down one at a time. How does this feel when you do this? How does your psychology change?

Please understand that when we use the word psychology, we are not talking about psychoanalysis, we are not talking about Freud or anything scientific. We are merely talking about your sense of your self. It is that simple, something has changed within you, and consequently you are changed. Just pay attention to the shifts.

Now that you have the sense that you can radiate light, try moving your limbs in such a way that you are throwing the light a great distance. This is a very energized movement, full of passion and commitment. You are connected to the world around you by the light that you are controlling.

Walk around and greet the people in this room. You don't have to do anything but that.

Now, let's throw the light again, but this time just toss it nearby, easily, tenderly.

Is there a difference from throwing it far?

Heat is another attribute connected to fire. Imagine that not only light is leaving you, but warmth also. Warmth is going out from your hand and you begin to warm the objects in the space.

This is not so difficult, but you have to accept that it is possible and then it is.

Something is leaving you and it is warm. This automatically makes you appealing to others. Go to people and give them some warmth because they need it. Go to a chair and sit, but warm it up with your tailbone first, then sit in it and feel your light and warmth going into the chair and down into the earth.

Radiate sitting, sit using the warmth and know what you have done.

Let's return to your monologue. Each word is a piece of light coming out of your mouth and lighting up the space between you and your partners, between you and the audience. It is a radiant activity, not like the forming activity of moulding.

Romeo makes Juliet beautiful and Juliet makes Romeo beautiful, because their love radiates towards each other. They light each other up. They light up and warm up the world and each other:

> But soft, what light through yonder window breaks?
> It is the east and Juliet is the sun.
> Arise, fair sun, and kill the envious moon,
> Who is already sick and pale with grief.

Shakespeare knows these things and he gives them to actors to use with the words they speak.

How was it to speak your monologue?

S: *It felt huge, powerful, and even vulnerable.*

Sounds like a good experience to me. How did you get there?

That is the work. The work brings you there. Do you have a clear sense of the process you used to feel that?

Clap: Start speaking the monologue with radiating, then change to moulding or flowing. You choose how you will speak, then back to radiating, etc. Just switch quality with my clapping.

S: *It is interesting what you can do.*

Yes, there are so many things you can do with the Chekhov technique.

It isn't about your own personal life. Of course it is you doing it, and that is unique and individual. It is the artist in you forming harnessing and expressing itself.

We know that love contains light and warmth, but it isn't that you are sending out love. This could easily become sentimental. It is about using the light and warmth, which are objective things, to express this archetypal idea of love

The light and heat could also be quite violent in an explosion, for example, a volcano.

What happens when you use this image of the volcano?

S: *I certainly felt hot, I mean in a passionate way.*

Okay. Now use this gentle image: In the night, in the middle of a forest is a small house, and in the window is a candle burning. It is still, light, warm.

S: *It was as if I could welcome anyone into my space. It was kind of respectful.*

Very different from the volcano, yet the same elements give you this or that impulse. The range is large when you begin to penetrate the surface and investigate. It is fun and quite easy to do, you just have to do it and it will happen. It is a way to practise, to just do it. Chekhov suggests this, he calls it 'continuous acting'. Give yourself a task it will take 3 or 4 or 15 minutes out of your day, but it is a way to practise. Go out into the world and have a simple interaction. Walk to the store and buy some bread, for example, anything. Choose a quality to explore. Begin as you leave your home and end it when you return.

Let's work on the remaining element, air.

It is weightless and thin, nothing can really take form in it. It is very far from earth in quality. It is quick and light. Chekhov gave this quality the name Flying. It's as though the gestures you make, the actions you do, will not remain with you, but they fly off into the space, then disappear. It is like trying to catch a bus that is leaving before you get there and you are trying to fly to it, so you can get on it.

Very quickly turn your head to the left, the right, up and down, look behind you all as if there was a wasp tormenting you. It stays nearby always around your head.

Walk as if you are on very hot sand and you have to get to the other side of the space. These are naturally occurring conditions for flying. You have already done them; they are a kind of flying.

If you know you are flying, then you can control it so it appears as a panic or chaos, but it is art instead of chaos.

Things are flying away from you. It is like blowing dust. Make a movement like that. It leaves, is gone and it's as if there was never any form to it.

'What?' 'Who's there?'

Stand still and imagine that you are making these flying movements.

Now you will fly the monologue. Go.

Stop. That was very loud.

Do quick and light translate as loud? Try to work with a feeling of ease now. Do it easily and a little more quietly. It isn't about shouting, it is about speaking words so quickly that one wonders if they were ever said, or who said them. 'Did I say that? Did I say that?'

Okay. I will clap my hands again and you will change from flying to any one of the other qualities you choose to do at the moment of changing.

This was crazy, right?

How was this for you?

S: 1 Freeing.

2 It is like playing, like being a superhero as a kid.

3 Very difficult to keep up with your claps. It was too quick, I had to think about it and so I couldn't really do it. I lost the lines.

Well, that's it right there. You cannot really do it if you are thinking about it. You have to know that you can do it and then you don't have to think about it. It is like picking up this bottle of water, I know I can do that and so it requires no thought, only the will to do it. So you have to practise *how* to do it, so you will know that you can do it.

But playing with these qualities of movement breaks the lock you may have put on the script. It allows you to channel the energy specifically rather than being all over the place . . . These four elements are the building blocks of the universe . . . the *archetypes* of *how*, so to speak.

S: *This is something that really seems to change the energy in the space.*

Can you feel that?

This principle of quality is at your disposal, use it to change the significance of what you are saying. It is working with a feeling of form. So now we have these tools to play with when examining the script.

On stage everything you do is significant. You have to make a choice about what you throw away or what you point up.

This approach applies very sweetly to Shakespeare. Often there are such long speeches and we have to find ways to make them work. Everything is a piece of art.

These four ways of moving have a strong impact, not only on the activities we must do, the business, etc., but also on our psychological gestures. We can approach these gestures with real formative powers when we apply the qualities. But you have to practise moving with the four qualities and become very familiar with how they work on you, how they change again and again.

Here again the archetype is leading us to the particular.

8 SENSATION

Today we need to pick up the *Feeling World* in the scene, so we will talk about and work with Sensations as the means to experiencing the feelings of the characters.

Everybody knows this experience, so it worth talking about. Do you remember one day going to sit in a chair, and you completely commit to the sitting, expecting the chair to be a certain height, but in fact it is a few centimetres lower than you expected? It is a quick sensation of falling and in the pit of your stomach there arises a frightful feeling. It is very brief, but nonetheless memorable. This is one of the primary sensations we will look at today. Falling.

So watch me, I will begin to fall, but I will catch myself with my foot and leg before it is too late, but you can see that I did have a moment of falling. My body knows this immediately and I experience a very uncomfortable sensation that stops when I catch myself.

Watch, I will do it again. Notice I am not bending at my waist but I am falling because I am bending from the ankles and I actually loose my footing, but I will have time to recover before I hit the floor. I don't want anyone to hurt themselves, but I would like for you to experience what falling is and what it means to your body, and what your body communicates to your feeling life.

S: *I felt very uncomfortable watching you in that brief moment. I felt as if it was happening to me.*

Good. It was happening to you. I did that to you. Whatever way you can feel it is all right. But you want to be able to experience yourself going down, falling down.

A dancer or an acrobat learns to fall without injury so the panic reaction fades, and we know this about them; we believe that they won't get hurt, so we do not have the same experience while watching them fall. In a normal person, the slightest inkling of a fall creates a panic within.

Fall forwards, fall backwards, fall to the side. They are all different. We are indeed looking to wake up the panic, even if it is brief. It is natural, it is a true psycho/physical thing. You have to catch yourself, otherwise you will get hurt, please do not hurt yourself.

Now fall into this chair. I will be standing here to stop it from going backwards when you do fall into it. Fall. It is not the same as sitting, allow yourself to fall down into the chair. If the panic does not come to you, then you have not fallen, you merely sat in the chair heavily . . . Can you feel that?

S: *Yes, that was really scary, I don't want to do it again.*

Yes, I felt you feeling it. No, you won't have to do that again.

Now fall backwards into the chair and try to do it just with the life-body. It is a strange request but I think you can do this now because you have already moved your life-bodies in other ways. We need now to make them fall down, and to keep them in a condition of falling down for as long as we want to. Because the sensation is so brief, we have to take that little moment as the sweet spot. It is intense and it is real. This is what we want to sustain. The very beginning of the fall is the sweet spot. Not the very edge of falling, but actually the moment the life-body begins falling. Perhaps you will resist this moment, because the sensation of falling is scary. It is natural to resist this because it is a primitive sensation, and you want to protect yourself. Try to get past it, you won't be physically falling, but 'inwardly falling'.

Keep your life-body falling.

If you fall all the way into the chair, then two things will happen:

1 The fall will end and so will the sensation.
2 Your life-body will be so far away from your physical body, you will lose contact with it and consequently nothing will happen. You will also have to start the falling all over again.

Your life-body can only be a few inches away from your own body. If you become too detached from the life-body, too far away, you've lost it.

Start the fall, keep falling, walk to another chair, let the fall bring you into the chair, then stop the falling.

S: *Is it the first shock you get, and you just stick with that?*

Yes, the first shock of it, but I caution you not to stick with that, because that already sounds to me as if the movement is stuck or stopped. It has to be a continual and a continuous event. It is an inner event, something that is happening now, not something that has happened and which you are holding on to. It is an event. It is a movement imagination, not a visual imagination. Please don't try to see yourself falling, but feel yourself to be falling. The life-body is falling.

Let's try something that may be a bit easier. It is simpler any way. Let's take this energy that is moving in a downward way and localize it here in the chest, right where you believe your heart to be. Let's say that your inner heart or your energy heart is falling. Just begin falling from there and then try to sustain it. Don't let it fall to the ground. When you do that, the energy is no longer connected to the heart place, but it is connected to the knees or feet. We want this particular event to be always connected to the heart. Walk as you are doing this, walk to another chair in the room and sit. After you sit, then you can stop the falling heart.

S: *That was intense, but I felt like a zombie, really cut off from the world.*

Yes, you looked like a zombie while you were doing it, but I could see that you were doing it. What is important right now is that you could do it. Now you need to say yes to it. We will fix the zombie thing in a moment.

S: *It was a very powerful feeling of loss or emptiness that comes as a result of loss. It was really devastating.*

And how do you feel now that you have stopped it?

S: *I'm okay, it is all gone, but it seems kind of mysterious that it happened. I mean, it seemed to come from nowhere.*

No, it came from the fall. It's really what happens to us. Here we are recreating that sensation. We say something like, 'I was really down'. Or ' He fell into despair'. We speak these phrases. Where do they come from? So, the heart falling, that is actually what is happening when you are down, devastated, heart broken, etc. Your spirit is down, or going down in some way when those things happen.

As actors using this technique, you have the power to start it and stop it when you need to. And when you stop it, there is no residue.

Nor do you have to spend time outside the rehearsal room thinking about the time when you were a child and your dog died, or whatever. You just make a connection to the sensation that comes to you because of falling downwards.

Essentially, negative things, thoughts, feelings, lack of will, have a tendency to go downwards for us.

This is knowledge that you can use . . . Pay attention to what is happening to you.

Let's see if we can fix the zombie effect. Concentration is critical in this work. But we have to learn to concentrate on one thing while doing another. When you are so concentrated that you lose contact with the world around you, then none of this juicy stuff is useful; it becomes a problem.

So start again with this falling heart and move around the room. When you sit in the chair you can stop it. But while you are moving around the room, take your time and really see what and who is around you. Find someone, and tell them this: 'I have a body and my body is expressive.' Speak out of the moment, speak the truth of how you feel right now. Stay awake and present. It is easier than you think. Remember it is not about thinking. You are just falling and the sensation of falling gives you a lot to play with. Say yes to the littlest thing you sense in your body; more will come.

This direction of down is really the important thing here. We can change its specific meaning, whether your heart falls, or you ears fall, or your eyes, or even your genitals. Down is important. This is work on the vertical line, the tragic line, if you will.

It is possible to work with just location and energy. For example, please imagine that energy is leaking out of your elbows. It is a kind of pathetic tragedy, a human being's vital energy leaking out of her elbows. Old cracks that are unfixable, and the energy leaks out in a downward way, dripping down wasted onto the earth.

What do you feel like now? Is there something here?

S: *It is incredible how weak and apathetic I just felt after I allowed that leaking to start.*

Research shows that the elbows have a correspondence to the will. If the energy is leaking from this place, then it is not surprising that you could feel apathy. Once we begin to investigate these things, we find such nuanced detailed ways to express ourselves. Our understanding of the moments in the play or the whole play itself can take on a more poetic quality for us. It is also just an enjoyable way to work and look at things.

Try to fall from the genitals now. We know that this is the will centre. What kind of will is awakened by this fall?

While you are doing that, look at someone and say, 'Let's go'.

What is that for you? Is this your normal state of affairs? How have you changed? What circumstances could you be in right now?

S: *It was the strangest thing.*

I wonder if you could express yourself differently now, and be more specific than saying it was strange. I know that what I asked you to do is strange to begin with, but to say it was strange is really to say nothing. Spy back and try to work it out so that your response is specific, then it will become very useful to you.

Saying it was strange is a way to dismiss it. You have to really plunge in and look at this stuff and then you will know it. Talk like artists, talk like actors.

S: *It was pathetic to feel this inability to do anything. And it was almost comical to say, 'Let's go', because I couldn't bring myself to move. I mean I could have moved if I really wanted to, but I experienced a kind of character or a moment in a character's life where she could not go on. It was thrilling to find this. And it was so easy.*

Okay. This is an actor talking. Perhaps next month you will come across a situation, or be involved in a rehearsal and you will be called upon to portray this very thing. You will know how to do it. And if it is the correct thing, you will know it. And if you choose to do that, it will always be a living thing, because it is present and active, not remembered. You will be able to do it again and again.

Can you split into two groups, half and half? Now this group here will fall from the heart and this group will watch.

Good. Thank you. Now my interest was in the watchers, the audience as it were. And my question is to you. What happened for you while you watched these actors falling?

S: *Your heart goes out to them.*

Because the actors are actually doing it, the fall is actually what happens. The audience also has life-bodies and the energy there begins to move when they witness the fall happening, and they themselves inwardly fall out of sympathy or empathy. Here we have a way to reach beyond the footlights, as it were, or reach beyond the fourth wall right to our audience.

There is no need to remember things, the body will take care of it. The source of your feeling life will become objective. The audience will be moved.

The trick is, don't try to feel sad. I know the scene calls for it. So, you have to fall. You have to actually fall in rehearsal, so you set it for yourself. In the performance, your body knows so well what to do, because it has already been doing it in rehearsal.

Now that you can follow the fall and respond to it, we will try something different. See if it is possible to resist the fall. This is real. It is probably the most real thing. Nobody wants to feel this stuff. Only actors want to feel big things. Regular people become drug addicts or alcoholics in order not to feel things. It is a curious thing to watch someone struggling not to feel.

Start the fall and keep it going, but resist it. Do anything else rather than go with it. But don't cut off the fall, you need to have that going so you have something to resist. I fall, NO. I fall, NO.

Continue in this way resisting for a while, and then say yes to the fall, let it take you where it wants to take you now.

I fall, YES. I fall, YES.

S: *When I resisted the fall and then let go, the fall was so much more powerful.*

The consequence was building, and then the dam breaks with a flood of feeling.

The opposite of falling, is floating. Now everything is going up, it is impossible to fall. You cannot fall. You will not fall.

Up, when I go up. It is impossible to sit in a chair. Everything keeps going up.

Now the life-body is levitating out in front of you, floating on its back. It is just floating on air. It is a ridiculous crazy image, but if you become one with the direction of upwardness, you will find the corresponding sensation. It keeps going up, just as it kept going down when we worked on falling.

Make your heart float up but not so high that it loses contact with the heart space in the body. By just standing still you can make so much movement happen and feel very alive because of it.

Imagine the inner brain is floating up. What happens when you do this?

S: *It made me giddy and silly. I laughed for no reason. Everybody seemed to be in agreement about some joke or something.*

Notice how it is so different from the sensation of falling. It moves in the opposite direction and it takes you far away from the despair of falling.

Float from the genitals and say 'Let's go'.

S: *What a difference, I wanted to go with such eagerness when I said those words, like we were going to a long awaited prize.*

This is the way to evaluate things. Circumstances are so important to actors. The inner event of this particular floating gave you some circumstance and makes it possible for you to act freely. This is how you can justify your practice.

We are actors, so in a normal situation, we allow the circumstances of the play to affect us. When the given circumstances are clear, they will speak to us. We will be involved in them and this falling or floating will give us the very specific feelings we need there.

The third primary sensation is balance or balancing, or finding the equilibrium and sustaining it. Balancing prevents you from falling or floating. What do you have to do to not fall? The tightrope walker is about to fall, and then, Ha! He catches himself. This is the moment of catching the balance.

What does that feel like?

What are you doing? What are you experiencing? What does this sensation do to you?

S: *Staying alive?*

Is that what you experienced? What does it take to stay alive? It's the moment of revelation. All of my resources come to bear upon the fact that I am going to stay. I am not going to fall. I understand now. I know.

See if you can find it and sustain it as an inner thing.

Now the head falls, catch it. Sustain the moment, the sensation of catching it.

S: *Is it too much to say that balance is a victorious sensation?*

I think it is sometimes, depending on the fall that precedes it, it can be a tenuous sensation. I think it can also cause a pause – uncertainty.

Learn to evaluate the outcome of your experiments. What parts worked, and what parts did not work? Then the technique will progress in exciting ways and you will never be lost.

9 FEELING LEADS TO ACTION

When you go to the seashore you find this seaweed that attaches itself to rock. When a wave comes in, the seaweed begins to float in the water. Feel yourself float up as if you are this seaweed. When the waves go out, there is nothing left to support you and you must fall. When the water comes back in, you float up. You are up when you are being supported and you are down when the water is out. You are not going down, you are falling because there is no support. Know the difference between going down or falling, then being down, and going upwards or floating, then being up.

If you feel something, you have the right to express it, and your body is the only thing you have to express that with. It is not enough to stand there and say the lines. Keep the rising going, or keep the falling going. You have a right to express what is going on inside of you, so do it. Try to connect this physical exercise with what you will be doing with your life-body; this floating up, or falling down. After the water goes out and you have fallen, say, 'I fall'. When the water comes in to support you and you are up, say, 'I rise'. Say it out loud as a fact, because it is so. Be involved 100 per cent in what you are doing.

Now stop the physical body from doing it and work only with the life-body.

I fall. I rise.

S: *I am having trouble visualizing the life-body's heart rising and falling. I don't know how to — the steps of it.*

As I said before, you want to experience this as an event; it is not visualization. It is actualization. It is a movement imagination.

To be able to visualize a character in a picture sense is one thing — but when we are acting, we have to incorporate the image. That means to put it in the body somehow. That is why we try to move the life-body so that these things can happen as movements. Unless you incorporate it, there is very little to express. How are you going to fill up? You will fill up by doing it, really doing it.

S: *I imagine the energy that I feel in the heart area — the energy that is concentrated in the chest area — that is what I make fall?*

That is right. Make that energy fall in a downward direction, or float it in an upward direction.

Now you will expand. Say, 'I grow'. Actually grow.

Speak the words only if they are true. Keep growing.

Now you will contract. You have to do these things first with the physical body, then use the artistic frame, and then use only the life-body. As you contract physically, say, 'I recede'. Feel what that means to you. Now you can express yourself, you have something to express. You have a right to say it, because it is happening. Let it into the body. Let it out of the body. I mean you experience it with the body and you also express it with the body. The body is the only thing you have. Keep receding, ride it for as long as you can.

Keep receding. Your heart is more distant, less in contact with what is out there, more isolated and alone. And if there is something there that you are feeling, let it out. Don't be afraid of it or ashamed of it, let it out. You are actors you have the right. Follow it, it is leading you somewhere – right where you want to go.

Express what is there. You have a right to do that – You have an obligation to do that. Don't stop the event. Don't stop the activity.

Now we have the option. I grow, NO. I grow, YES. You have the choice. So, you never stop the event, but you are fighting it now. We do this all the time. Something within us says, 'I'm not going there. Let's not go there.' So, if it's NO, what are you going to do about it?

Now you will work with a partner. Let's say the relationship between you is a sympathetic one: you love each other. Just stand there facing each other and someone will begin by falling or rising or growing or receding. When this happens and only if it is happening, then this person will say to the partner what is going on for him. So, for example, if you are falling, then say out loud 'I fall'. If it is actually happening, then your partner will see this happening and the partner will say, 'I watch you fall'. Because of the defined relationship, certain responses will begin to emerge and so all you have to do is follow them. Honour the relationship. It is not pleasant to watch a loved one falling. Because you love that person, perhaps you too will begin to fall in sympathy. If that is so then now you have the right to say 'I fall' and your partner will say, 'I watch you fall'. This exercise will move by itself, don't worry. The changes will come because of the relationship. Just follow it however it goes. Be sure to say, 'I watch you', so as to validate the experience of the other. You have to start it somewhere, but then let it follow its own organic course. Follow the inner movement rather than forcing a change. It is leading you somewhere. If you feel yourself starting to fall, then just follow it. Don't just do tricks now. Follow some natural thing. Where you are being led by the relationship? Use your body to express something; let it work for you.

Are you experiencing the relationship? Does it make you want to do something? We are dealing with the feeling life here. Does it make you feel like you want to do something?

S: *It is taking me. So much is going on between my partner and me. I feel a tremendous amount. I can actually feel these directions moving through me. I can see them moving in him as well.*

Great. It seems to be working for you.

Now let's change the relationship. Let's call it antipathetic; you hate each other. Do exactly the same things in the same way but tune into the new relationship. It is a very different thing to watch your enemy falling, or to watch them growing. To fall in the face of the enemy is very difficult. Work with it. Live in the relationship.

S: *I can see how important a point of view can be. How it orients me in the scene with my partner. I really enjoyed hating her just now. There were so many possibilities to work with.*

It's an interesting thing. You said you enjoyed hating her. You are really giving yourselves an opportunity to play in the feeling world. You can see how it operates, and you have an appreciation for it.

Out of this world, we can take action. I feel this way, and because of it I want to do something.

So now try to stay in the feelings. Let them motivate your actions. When you feel like you want to do something – when you have the need to do something, then choose one of the six actions we have identified – I want. I reject. I give. I take. I hold my ground. I yield. Work with them as inner gestures. Don't just say the words. You have to do the gesture and then you can speak the words and this will be the truth, the affirmation of the movement. Make the inner gesture and as you are doing it, you are saying it. As you do it, you will follow impulses that arise from the gesture. Everything is coming off the feelings now. It is another way of looking at things. If we just work with the action, it is too cold. That is why we have led up to it this way. These feelings make you want to do something? So if you feel the need to take, then make the gesture and say, 'I take'. When you are working with the action, you do not need to say 'I watch you take'. Let's leave this response only in the feeling world. Keep this antipathetic relationship and begin again but now include the action. The willing world.

S: *I feel like I was saying NO to everything we did tonight.*

You don't trust it. You are saying to yourself, I should cry. I should have an intense emotion. It should be intense like it was the last time I did that. Don't do this to yourself. What you are saying is, 'I cry'. Don't. Just fall. Trust it, and follow the fall. 'I fall', that is what you should say. That is what you should do. You have to practise these things.

You have to have an awareness of what is going on for you. You can sense that some part of your being did this. Begin by looking around you. Watch someone crying in real life. Really look at someone crying in real life. You will see that they are in despair and that they are, in fact, falling inwardly. See that the spirit is sinking or falling. That is the best thing for you to do for acquiring this technique. You feel it here, and you see it affirmed in the world around you. Pay attention to the world and look for the objective human truth. Try to see the things that hold us together as humans. This is the work that we need to do. We have a tendency to look at things in such a personal way. This separates us from each other. What are the uniting elements?

S: *I think I get this. I mean, I understand the falling in terms of despair. The words 'he fell into despair' seem to me to be true. But we also say 'I fell in love' and this seems to me to be a positive thing, something we all desire to feel. Is this a fall? Do we fall in love?*

I think we must do that, because it is such an old way of expressing that condition. But maybe it isn't such a positive thing. Of course, it feels good to be in love. But it is still falling and it puts us on edge. I have had the same question and I have looked into it. We were playing at the studio one day with the four elements, falling into fire or water or earth. When we came to air and I started to fall into air, I realized there was no bottom to it, I could keep falling and never hit the ground because it was only air, and I could surrender to it, then the experience became quite beautiful and energizing. I became free by surrendering to the fall and I felt so giddy and large and soft and goofy, just like being in love. I think I now know what it means objectively to fall in love. To recreate that as actors, you have to give in to the fall. Here the greatest possible thing is to surrender to this fall that will never end.

Like falling asleep. We actually do fall into sleep, and you can observe that on the subway. It is a fall. All the idioms are real pictures of the movement. They describe what is true, but we have stopped having a real connection to these words, we merely say them without actually experiencing them. We live so much in our heads these days. What happens when you fall for the joke? What do you do with your body once you realized you were made a fool of in that way? Actually it is quite physical. Try it now, let's say you were conned. Have those guys doing the shell game ever played you, or the 3 Card Monty on the street? Watch the losers and what they do with their bodies. The world is really fantastic when we pay attention in this way. If you are an actor, it is good when you enter into this kind of relationship with the world. You see what is really happening, and it confirms your own experiences.

10 *DESIRE UNDER THE ELMS*

Let's return to the scene. Have you visualized what is going on? Have you imagined the scene played in a brilliant way? See it acted out to its fullest. First you see it, then you do it. It has to come out of something, so see it first.

The gesture is the action. Do the action and something will come of it.

If it is to be successful then we must look at the scene and see what the events are, or what must happen. We can play with these tools and then set the necessary moments in the body. We should be able to identify them and fix them somehow with the body. This does not preclude any idea of improvisation. We have to appreciate the form. Each time we do it, can be different but certain ideas, certain feelings are necessary to tell the story properly. They can be set in the body by using gestures and sensations. We will use the language we have developed in the class. In this rehearsal we won't use the script. Let's just try and see what happens between these two characters. We will use this language of action and reaction. Try it.

Her: *I grow.*
Him: *I hold my ground.*
Her: *I grow.*

Forget the script. It is an improvisation. You are thinking too much about the words of the play and you are not paying attention to her. Use her. She has grown twice and you haven't watched it once. In lieu of table work, we could do this improvisation and learn so much about the scene.

Her: *I grow.*
Him: *Hold ground.*
Her: *I grow.*
Him: *I recede.*

Again you did not see her grow, I *watch* you grow. This is a game. We *play* through the scene as opposed to being scientists with pencils saying, I think this would make me feel like this. And then refer to our notes later. This allows you to act and bypass your intellect for the time being.

SECOND SCENE: Abby and Eben

Her: I grow.
Him: I watch you grow.
Her: I grow.
Him: I watch you grow.
Her: I grow.
Him: I hold my ground.
Her: I take.
Him: I recede.
Her: I watch you recede.
Him: I reject.
Her: I take.
Him: I rise.

Stop. Were you rising?

S: Yes I think so.

No. I didn't see it. I want to point out that what happened
there was quite natural. She said something that confused you,
and you were falling from the head. You were grappling with
that. 'I fall' is the thing to say there, because that is what you
are doing, and then she is going to take some more. You have
to stay alive in this. How does he feel about her? Not what he
does, how he feels. He is drawn in against his will, and how
does that make him feel. It will activate you to another place, if
you go with it and admit it and realize how you feel.

Try different things. Don't be afraid to act. Feel it first and go
from there. It is just a little game to play as actors. Instead of
having a discussion about it, we will work this way tonight.

Let's see another couple:

Her: I grow.
Him: I watch you grow.
Her: I grow.
Him: I watch you grow.

Her: *I take.*
Him: *I fall.*
Her: *I watch you fall.*
Him: *I fall.*
Her: *I rise.*
Him: *I watch you rise.*
Him: *I recede.*
Her: *I watch you recede.*
Him: *I reject.*
Her: *I hold my ground.*
Him: *I fall.*
Her: *I watch you fall.*
Him: *I reject.*
Her: *I rise.*
Him: *I watch you rise.*
Her: *I grow.*
Him: *I watch you grow.*
Him: *I recede.*
Her: *I watch you recede.*
Her: *I rise.*
Him: *I watch you rise.*
Her: *I rise.*

Good. To do this you need to see the scene acted out first in your imagination. That way you see the movement and you can do this improvisation. Once you have this basic understanding, then all you need to do is insert the playwright's words.

Let's clarify our vocabulary. 'I rise' is to go up. 'I grow' is to become powerful. Expanding spherically is to grow, moving out in all directions.

Rising, to rise upward – ascending – floating.

This is a warm approach to acting. Sitting at a table with a pencil is cold. This gets you closer to acting. You have to know when you are giving or taking. If you can visualize the scene, then you won't be conflicted about your choice. It can only be one thing.

Once we understand and accept the laws of the universe, we are no longer slaves to them. We can manipulate the laws and use them.

We tend to complicate things because we believe that our lives are so complicated – we are just subjected to very basic laws of movement. Once we can get these tools, we find that they are objective, and when we use them they set us free.

We don't have to bring evil home to our families, even if we are working with evil themes in the play. We can drop it when we are finished rehearsing for the day or finished performing. We have to, otherwise we are on a road to madness.

Do what you feel is going on, and then say what is going on. This is just an early rehearsal. Try to play as opposed to work. Find the joy through this approach. We are able to find very serious things through play. We used to be called players. We must find an element of play in the work. This is a game here. Through the game we can discover some moments that we need in the scene. You may discover a seed for the whole scene.

You should know the scene. You had to have made some agreements about it. The point is to get information that is helpful when you work on the scene through the game. You are creating signposts, or anchors for when you do go through the scene.

You don't have to do everything you are taught. You get a lot of rules, a lot of tools. You only have to use what you need. Sometimes, you don't need anything. A technique is useful in the times you feel deficient on some level and you need to fall back on something. Otherwise you are an inspired actor. You just act on inspiration.

FIRST SCENE:
Her: *I rise.*
Him: *I watch you rise – I reject, I grow.*

Her: I *watch you grow* − I *fall*, NO − I *hold my ground* − I *grow*.
Him: I *watch you grow* − I *reject*.
Her: I *give*.
Him: I *fall*, NO.
Her: I *watch you fall*, NO, I *rise*.
Him: I *watch you rise* − I *reject*, I *reject* − I *grow*.
Her: I *watch you grow* − I *rise*.
Him: I *watch you rise* − I *reject*.
Her: I *give* − I *grow*.
Him: I *watch you grow* − I *fall*, NO.
Her: I *watch you fall*, NO − I *rise*.
Him: I *watch you rise*.

Okay, something is happening here. I think you are getting it now; this was a bit more complex and full. Feeling is great for actors. But the actions are going to give us the behaviour. Don't go straight to the actions, because that is cold. Go to the feeling life first, in order to find the actions.

Chekhov is adamant about this: the intellect is the killer of art. Analysis is intellectual. When you begin with analysis, you are not using your actor's soul, you are using your intellect.

If you do the gesture (the action) in rehearsal and speak the truth from that inner action, then when you are in performance your body will remember that gesture and it will be there for you. Behaviour will come and it will be from the inner movement.

S: *Can you explain the 'I am' for me? I don't know the 'I am'.*

Maybe you don't know who he is. 'I am' is the gesture for the character. I am = neither feeling nor action. It is the statement of the character, it is the archetype.

There are variables here. You don't want to make this work so cut and dry. Expansion and contraction are dynamic principles. Expanding is both positive and negative. Contraction is both positive and negative. Don't limit it to just one or the other. Explore the possibilities and play around with them. See if you can find an 'I am' moment. When the character is just saying 'I am' and nothing more. There are plenty of these moments in the play. This is what we need in rehearsal, to know the scene, the events of the scene, the dance of the scene. Then we can easily learn our lines.

What does she want?

S: *To make a connection with him; to win him over.*

Fine. How do you win someone over? Talk to me and use your hands as you speak. This will lead you to the gesture.

S: *Open up to him.*

Great! You see how she is giving in her gesture here? She is opening and this gesture is a giving of herself. Now speak the lines and open, give to him as an inner gesture. You want to find one gesture that will work for you, not one that will change throughout the scene on each line. Simple things are the best. You can change the rhythm of simple things, and still hold on to the simple image. We want simple ideas with complicated rhythms.

Open up to him. Then, if you follow the inner movement, if it is the real thing, an impulse will arise and behaviour will come out of it. Something will come of it.

What if she is trying to bridge the gap? Warm up the air between them?

11 THINKING

Let's look at another way of using the technique, let's consider the thinking world.

Using the technique can give freedom from never having to think again on the stage. This may come as a relief to some of you, or perhaps scare some others.

Chekhov said that if we really did think on stage the play would come to a halt because it takes too much time to really think. The play is happening in theatrical time but to really think involves another sense of time. So we must *perform* thinking. Actors are horrified to think that, but why not try it.

What is important is that the audience knows what the things you are thinking about mean to you. Everybody knows that we think in the head. Our brain is there and we even feel ourselves thinking from our heads. So if we apply a specific inner activity and locate it here in the head, we will appear to be thinking because the audience will see that this part of the body is energized in activity. What the thought means to me is of more value than the thought itself. Since no one can really know what I am thinking, then it doesn't matter what is happening in my head so long as it is conveying the necessary meaning. It is much easier to do it and to see it than to talk about it. So let's try to do it.

Expand the inner left eye. It is not about having a large eye, but about the activity of expanding just here around this inner eye. Keep this going, put all your concentration on the activity.

What does it feel like is going on for you with this? Can you do it?

Now let's see what happens. Will you three over here do the activity? Go ahead and concentrate on it, and we will watch you.

S: *It is fantastic, it really does look very active around the head and it is specific. Just like a specific thought. It isn't what I thought performing thinking would look like. I thought it was going to look fake and indicated, but they were very involved with something.*

That's right, it isn't about 'showing' that you are thinking, it's about living in the quality of a particular thought and reacting to that.

Now you three try it and the rest will watch. After a time, and I trust you will know when, say these words, 'Of course I knew this all along.' And then stop the activity.

S: It tells a little story. We know that she was pleased with herself. She was just placed instantly into some kind of circumstance.

S: It felt that way. When I was doing it, I could feel a world around me and I could go into it and live freely in it. But it was short-lived.

That was the actor in you coming alive, wanting to act. We are not playwrights. We are actors who will act the given circumstances. We have to conjure up the circumstances that have been given to us in order to enter them so that we can live in them and relate to them.

Everybody, please contract the left eye and say, when you feel it, 'Huh, I hadn't considered that'.

Now this is something completely different, but from the point of view of the technique it is the same, a simple activity located somewhere in the head. But this one has a different meaning and a different reaction.

If you are actually engaged in something, like an expansion or contraction in the area of the head, then it will be as if you are thinking. Doing this is more interesting and enjoyable then trying to think or pretending to think. Ultimately you can only really think up to a point about something. I mean, you the actor will have thought this out in rehearsal already, so you will in the end be pretending to think on the stage, because there is nothing more to think about. It isn't an action like giving or rejecting, which you can really do again and again; thinking is something very different from an action. But it is an activity. *Performing thinking is an activity.* Your chosen activity is happening in the present and you are reacting to it, so there will be a truth in it. It is up to you to choose the truth you need at a specific moment.

Please try to have your inner ears fall. They keep falling down, but remember they don't get too far away from their place. Use downward moving energy just here by the ears. Give yourself over to the activity and let the actor in you wake up. This is another thought. I can see that, just by looking at all of you. It is wonderful to see *that you can do* it. Please say out loud 'I can do that'.

S: *I can do that!*

So you have told yourself you could do it and now there won't be any doubts; it is yours and you can do it.

You need to know what each different activity means to you. Then it will easily translate into what the meaning of the thought could be.

You can look to the six directions again as a way to *practise* different inner activities, and it will soon become clear where you are with this and how you can make it work for you.

Inwardly move the head in the direction of up . . . say, 'what a fantastic idea'.

Now let's try this sequence:

1   Head goes up . . . say, 'What a fantastic idea.'
2   The left eye contracts . . . say, 'But I hadn't thought about that part of it.'
3   The ears falling down . . . say, 'We're in big trouble now.'

Whatever you can make happen in the area of the head will give you something to react to. You have to play around with this and you can have some good fun discovering what this activity and that activity means. Speak whatever the activity suggests to you, only know that you are involved with 'thinking' and that you are *performing* thinking as you do this.

12 ARCHETYPE: THE PSYCHOLOGICAL GESTURE

The psychological gesture is an archetypal thing in itself. It is very large, it takes up a lot of space, and it holds together a unifying idea. The gesture creates streams of energy in the body, and what comes out of the body is the intention of the character.

Tonight we are going to break it down just for the character. We are going to use the psychological gesture to help find the character. We can find our way to a specific character by using a very large idea, a primitive and energetic image. The archetype is the biggest possible idea we can find for the character.

Archetypes are prototypes. As images they vibrate. The energy in the image vibrates within us.

You have become sensitized to energy and inner movement through other exercises. Pay attention now to that. Please notice that when we face archetypes, we experience an impulse immediately. Look for this impulse in the six directions of up and down, forwards and backwards, expansion and contraction. Stand there quietly, wait for an impulse to move in one of these directions, I will name some archetypes and you will softly repeat the word and an impulse will come to you to move in a direction. Just point up if it is up, or point forwards if it is forwards, etc., honour the direction by acknowledging it:

— The hero: which direction does it move in you?
— The king: which direction does it move in you?
— The orphan: which direction does it move in you?
— The coward: which direction does it move in you?

Honour it by stepping in the direction of your impulsive response:

— The prince.
— The loser.
— The virgin.
— The gambler.
— The mother.
— The wizard.
— The warrior.
— The traitor.
— The actor.

Was there a connection there? Did you feel the impulses?

S: *Yes, I felt something each time you gave us a new image. Some were mixed in the direction like up and forwards. Is that OK?*

Yes. It is what it is. When you put all your attention on the directions, it is an easy way to catch the impulse, and connect with the energy that is in the image. It becomes movement again, so it will speak to you. You can discover many things quickly about the character. It will hold the character together, because the archetype is a whole and complete energetic thing.

Beware! We never want to play the archetype. It is the model we draw from. It is the food we eat as actors. We don't present the archetype, it is too powerful, too pure, too big. Certain plays may call for archetypal characters and so it can be great fun to express yourself in that way. It's a question of style. But normally the audience is quickly overwhelmed by the force of it, and then it's over. You've 'blown your wad', so to speak. Please remember this. You want to play a character who is a particular human being in a particular circumstance. The archetype is powerful, but also general, so we have to be cautious in approaching it and expressing it.

When you are looking for the character's archetype, you are looking for the archetype, not the character.

The archetype is not the character, the archetype is only the will force in the character . . . Having found the direction that the archetype moves is the first step in finding the psychological gesture for that archetype. Already you know the gesture will move in a certain direction, it follows the impulse. This is most likely 85 per cent of the gesture. The remaining 15 per cent is in forming the essential image. There is very little thinking here, as you will see.

Stand in a circle, but turn out and away from each other so no will influence your movement. You will be on your own to invent this gesture. I will name an archetype and you will softly repeat it, then you will know in which direction it is moving for you. After this I will say '1–2–3 go' and you will make a gesture that expresses this image. It is your gesture. You will find it and you will commit to it. Everybody ready?

The witch.

1–2–3 go.

Are you happy with your gesture?

Are you awakened by this gesture? Is it giving you something? If not, abandon it immediately.

Change it. Return to the impulse or the direction. You can trust that the impulse is correct. Chekhov said the impulse of the artist is always correct. You can trust in that. Perhaps the image is incorrect, or the gesture could be better. If you like it, then make it better and play with it, commit to the gesture.

What we really want is the gesture, so we are naming the archetype to help find the gesture.

Let's look at them now. Make a circle and everybody will face in so we can see each other. 1–2–3 go . . . Look around at what has been done.

It's very interesting when we see that they are all more or less the same – the gesture of the witch. It lives in us. We recognize it as a form. It is universal. This is how the archetype works on us. Yes, they are all a little bit different. This is evidence of your individuality. But they are basically the same, the direction, the use of the hands and the head. So you see finding the gesture is not so difficult, it is basically given to us and it does excite us in some way.

The point of working with the archetype is to experience the will force of the archetype. The character will have this quality of will.

This is an expression/understanding of the will force.

The gesture leads us to the archetype, the archetype also leads us to the gesture.

Make the gesture three times in space, radiate it, look at it with your new eyes in your shoulder blades. Say, 'I am'.

Make the gesture an nner movement now. Follow impulses from it.

This psychological gesture is the crystallization of the will force.

Am I awakened by this gesture?

Is it giving me something? If not, abandon it immediately.

This is not the character. It is the archetype.

This psychological gesture is the expression of the essence of this archetype.

Let's find the psychological gesture for the gambler.

What is the fundamental, essential thing about a gambler?

S: *First thing that comes into my head is a table with cards.*

You are now making a story. It is not a narrative or a story and you are not acting anything out.

S: *The gambler wants money, he wants things to come in to him*

We all want money, we all want things to come to us. The gesture of the gambler is a bit more special and specific than that. Try the gesture of 'I want'. Is it correct?

Greed is not it.

I am sure that you will get a lot from each of those gestures, but you are seeking the will force of the gambler, not the miser. Although some are greedy, not every gambler is necessarily greedy. The essence of the gambler is risk taking. What is the essence of risk taking? We are talking about the archetype, the highest possible form, not you as the gambler. If your gesture expresses greed, then it does not express risk. The purpose of this gesture for the gambler is to create an excitation in you for this risk taking. The gesture says, 'I am'. How can you make a gesture that moves forwards and excites risk in you.

The body is making the gesture, so the risk must be experienced in the body. Perhaps if you ended the gesture on one foot, with the rest of you hanging over the precipice, you may experience this risk. When you find it, you will know it because you will feel it. A particular quality of will is awakened by it.

S: *What does it mean — quality of will?*

It means how you do the things you do.
Here is another archetype: the slave.
Ask yourself. What is essential about the slave?

S: 1 *Lack of freedom.*
   2 *Oppressed.*
   3 *Struggle.*
   4 *Captive.*

Yes, all true, but what is the essence?

The essence is: the slave serves against his will. The undeniable essential thing is that. This is different from the servant who willingly serves.

You must ask that question all the time, ask what the essence is.

Stay on your feet as you explore your gestures. We are working with human forms. As we've already discussed, a big difference between you and an animal is that you are standing up on two feet. It is also easier to walk away from the gesture if you are standing up. You can walk away and look back at the gesture with your eyes in your shoulder blades. Leave the gesture but take the vibration with you. It becomes direct knowledge. Real knowledge comes into the body.

What is the essence of the warrior?

S: 1 *Fearless.*
   2 *Strength.*
   3 *Brave, courageous.*

These are qualities. They are not the essence. The father is brave and strong. The hero is brave and strong. The whore is also brave and strong. Look at it in another way. Try to find what is active in it. The archetype has to do with action, it is about the will of the character, the very specific quality of will. What does the warrior do? The warrior is the one who . . . what?

S: *Is it the one who fights?*

Yes, of course, it is the simplest of things. It is better if it is simple. You can keep this as a motto: 'Simple is best'.

S: *How do you choose the archetype for the character?*

Through deeds done. Aristotle said, 'a man's character is the sum total of his actions'. So, read the play you are in and write down the deeds done by the character in the play. It isn't really an analysis, it is an assembling of facts given by the playwright. Once you have the list, you can draw a line through it making a thread that connects these deeds. Then ask yourself who does this? What archetype does all of these deeds? It is a deductive activity, but it doesn't take long. Just by asking a few questions, we can know a lot.

In a sense, the gesture turns you into a magnet so all the things that belong to the world of the character stick to you. Now little things in the world around you have something to do specifically with your work. The gesture holds things together. The correct gesture is always moving closer to the character.

The character has no idea that he is being led by an archetype. Each of you is being led by an archetype, but you don't know what that is. It's like this: One day in your life you made a choice, and when it was over you saw that it was not the best choice you could have made, so you tell yourself that you will never do that again. Sometime later, in a similar situation, you repeat that choice and you see it was not the best choice and you tell yourself you won't do it again, but then later you do it. This is your archetype leading you. It is a kind of imprint that you follow, but you are not conscious of it. It directs your will.

Of course, sometimes these choices can be great and perfect choices. They will be consistent with who you are.

What the character wants has nothing to do with the archetype. It is what the character does. I am sure that Hitler wanted to save the world. He may have attached himself to an archetype, but that is not the archetype that led him, or the archetype that reveals him.

Let's look at Romeo from *Romeo and Juliet*. What do you think his archetype could be?

S: *The lover?*

Why? Because he is in love with Juliet? Because she loves him? Maybe this is true, but we have to look at his deeds and as soon as we do this we step into the play in a rich way, because we begin to work with patterns and we have overviews. The feeling of the whole becomes engaged. So let's go over his deeds and see what they reveal. Romeo's deeds done go something like this:

- He complains about love.
- He goes to the party — crashes it actually — a risky doing.
- Goes nuts for a girl, finds out she is a Capulet.
- He pursues it anyway. He returns to her place.
- He climbs over the wall of her house.
- He climbs another wall and professes love for her, kisses her.
- He meets with her nurse, says his intentions are honourable.
- He arranges their marriage.
- He gets married.
- He kills her cousin.
- He makes love to her.
- He leaves (banished).
- He returns under banishment.
- He kills Parris.
- He sees that Juliet's 'dead'.
- He then kills himself.

You see, these are the actual things done by him in the play.

And now the thread you draw through these deeds will be the archetype. The question to ask now is what kind of person does all these things?

S: *An impetuous person would do those things.*

Yes, this is true, but again this is a quality, not an archetype. On its own, impetuous doesn't vibrate. But the quality is useful because you can ask the question: what kinds of people are impetuous?

S: *Children are like that.*

But do you feel that he is a child? No.

You can rely on your gut, which says no here, so you must go on.

S: *A fool acts impetuously.*

A fool. How does that feel? Pretty good?

Look at the deeds. Is it a fool that does these things? It does seem that a fool would do these things.

Is he is a fool? Perhaps gambler? Perhaps rebel?

You do the psychological gesture that tastes best to you.

The fool seems to be most consistent with the deeds.

The fool's very nature is youthful and impetuous. 'Fools rush in where wise men fear to tread.' The fool is the one making mistakes.

S: *How can I act the archetype of the fool without knowing he is a fool.*

A fool doesn't know he's a fool. But anyway, we don't want to act the archetype. It quickly becomes boring to watch the archetype on stage, because there is something slightly unreal about it. We want to feed off of the archetype. How the will is affected, that is what we want. It is the energy and how it plays through us that is valuable. The archetype is just how the character is being driven.

If you analyse it with your intellect, it dampens the power.

Make the psychological gesture of the fool. Don't just walk away from it. Let it fill you. Take up space when you make the gesture. You are looking to take on the quality of will that lives in this archetype.

You can do this psychological gesture with the life-body alone. You don't have to be afraid of it, because it is only a force. We want this force. It is not a character. Say yes in a full way to the force that the gesture provides.

Is there something living in you?

S: *It is very seductive.*

Does this gesture excite me in the direction I want to be excited in? That is the first question to ask. If not then abandon it, and go on to something else. Move the body, take up as much space as possible.

This is not something to be analysed, it just is. That's what we can say about the archetype, 'It is.'

I was teaching an actor once. He is an African American. When I brought up the image of the slave, he just fell apart and was weeping on the floor. It was a very powerful image that had a very particular resonance with this young man. His expression was intense, and it took a lot of time for him to recover. He had trouble saying, 'It is.' Slavery is reprehensible and it makes perfect sense that he would feel this strongly, but he failed to see the archetype for what it is. It is the one who serves against his will. Hamlet is a slave, serving his father. He has to do this thing but he has no will to do it. In the play he is a prince but in his will he is a slave. Isn't that interesting? This is how I would do it if I was directing or playing the role. If we do not judge the word or the impulse, we are free to work with it.

Turn away from each other again. Let's try another one.

The hero — 1–2–3 go.

Use as much body as you can . . . move as a unit, but all the parts of the body move. Remember the artistic frame. Repeat this gesture three times once you find it.

Create the gesture so that it is a beautiful thing for you.

How is this? Is it different from the witch? Is there something in this for you?

Who is the hero?

S: *A person who saves.*

That is the saviour

S: *The fixer?*

That is the fixer.

S: *The courageous one.*

Come on. The one who is courageous, etc. defines a quality . . . keep it active. The hero is the one that takes up the quest. When we look culturally or historically at it, we see who we call a hero. We see that it is someone who says, 'I can't do this.' He meets a teacher who shows him the way. Then he says 'I can do this. I will take this on.' It doesn't have to be successful. It is the quality of will.

Let's look at these gestures of the hero. 1–2–3 go . . . You see again the overall similarities. The impulse is to go forwards and up with the body. The arms and hands and heads, even the legs are working towards this. This is the collective and unconscious agreement concerning this archetype.

S: *I just don't like the gesture yet.*

You think it could give you a bigger kick?

That is okay. I see that you are moving forwards and up, that is definitely the place to start from.

Now say to yourselves, 'I want to experience the sensation of defeat.' This simple command will begin some kind of fall within you. Once the fall is happening, make the gesture of the hero and you will see what happens to you.

With that sensation moving in the body, make your gesture. Radiate. Walk away. Look at it with your new eyes. Repeat it.

With the sensation of defeat, the inner movement is *down*, but the action of the *archetype* is *forwards and up*. With this sensation, the hero archetype becomes more specific and it gets closer to a character. We can combine these things that intellectually don't go together. Perhaps in the play we discover the will of the character is that of a hero but this character is a failure or completely reluctant to take up his quest or something along these lines. We can easily come to this idea by working through this process. We can play complicated things that have come to us in a simple manner.

In the book, *Lessons for the Professional Actor*, Chekhov says the psychological gesture and the archetype are the same thing.

We are really interested in the gesture more than the archetype. The psychological gesture is the 'flower of the technique'. It is a multifaceted tool. It can reveal for you a moment, or a scene, an action, a character, or the whole play. It is a way of synthesizing things. This gesture of the archetype says 'I am', while it is an amalgamation of the deeds done.

Chekhov says you don't need all these other things. Other techniques teach you to analyse, which means taking something apart. Then you have many pieces you have to contend with. Chekhov's technique helps us make a synthesis, it pulls things together, makes one thing out of a multitude. Some people are very seduced by analysis, 'wow this is a lot of good work and it is so interesting, I am working with my brain and so I feel very involved'. But really you are not physically active, nor are you acting. The Chekhov technique is simpler than everything else, but it is difficult in the beginning to settle the mind down and allow the simplicity to have its effect.

Let's take a look at this for a minute. Make this gesture of opening, with your face looking up towards the sky and the palms of the hands also facing upwards. How does it feel to make this gesture? Now do the opposite, open up with face looking down and hands facing towards the earth, bent a bit forward. What is the difference here?

The face, the chest, and the hands could be called organs of receptivity.

If we are receiving things here with everything facing upwards, then we are receiving from above. Culturally we have come to believe that heaven, and what is good and right, God, etc. come from above. We also believe the Devil, evil, hell, etc., come to us from below. It doesn't matter whether you believe in God or you are an atheist or whatever. You still have this physical and cultural understanding. It is very old and quite fixed within us. So when we open the hands this way or that way, place our chest and faces here or there, we can expect to receive certain things. We can really control the outcome of our gestures and use this knowledge to help us from floundering. We can go directly to what we seek.

S: *Is there a definitive list of archetypes?*

I don't know, but we can find them in mythological characters from Greece and Rome, Bible characters, the Tarot, the Zodiac, Grimm Brothers' Fairy Tales, the Mahabarata, folk stories from Africa or native America; all these are filled with the archetypes, those are the sources. They come from the old literature and arcane sciences that have survived the millennia. The reason they survived is because of their vibrational power. The things that do not vibrate in this way will fade away. And the archetype that vibrates in this way vibrates in all of us.

S: *Just listening to all this. I wasn't really thinking so hard about Eben, but I got this flash that his archetype is the rebel. Is that an archetype?*

Yes it is. Antigone is a rebel. And the way it came to you, in a flash, is a good sign. It came not as a rational thing but more as an intuition. Now that you have this idea, how does it sit with you? We are looking for a kind of inner bell to ring. If it feels flat, then move on. We want to be excited by our choices. But the rebel sounds good for Eben. He is always working to change how things should be perceived or accepted. He works against the status quo from the beginning. It seems to me like a good choice, at least a good place to start. You see how easy it was to find it. You only have to present the question to yourself and the answer will eventually come.

Since we are now here, does anyone have an image for Abby?

S: *She comes to the house and takes over all the duties expected of her and she takes care of these two men. Could she be the mother?*

Let me ask you right off, does it cause a bell to ring within you? How does that image resonate with you, who is going to play Abby?

S: *No bell. I was just guessing.*

S: *I don't think she is the mother, I mean she kills the baby, that is one of her deeds, that seems the antithesis of being a mother. That is like Medea who is no real mother.*

You have to look at her deeds. But I agree that the mother is not it. What does she do?

S: *Well she comes there with the intention of having everything, the farm, the money, the man, even the son. She can only get it by taking it and it doesn't belong to her.*

So who does those things? What kind of person does that kind of thing?

S: *A thief. That must be an archetype, it is so large and I feel it vibrating just saying it.*

Is there a bell? A shock? Is there a desire in you to work with the image?

S: *Yes.*

Then I say you have found a place to begin an investigation. You do not have to settle on that archetype ultimately. You may change it as your understanding of the material grows, which will happen in the course of rehearsing, but at least you have a starting point, a way of looking at it that is neither analytic nor personal, but you feel you know something about this. Try it. See what happens.

As soon as we add a centre, we get closer to a character. Go back to the physical gesture. Find the gesture for these two archetypes of the rebel and the thief.

Now make the same gesture for the archetype, but begin to move the gesture from a specific place – chest, pelvis, head, etc. You can use the imaginary centre and the imaginary body to create a vessel for the will force of the archetype to pour into.

Every new piece of information you add to the mix will more and more particularize the archetype into a character.

We have already said that Abby is a thinking type. Make the gesture of the thief with this quality of stick you found in your previous investigation on her as a thinking type. This is not wasted work; it will re-inform the gesture you found for the archetype.

13 IMAGINARY BODY

There is a story about how Stanislavski and Chekhov met one day in a café in Berlin a few years after Chekhov had emigrated from Russia. They had a long conversation concerning character, and they did not agree on how an actor should approach it. Stanislavski said that the character is in front of the actor and that the actor should pull this character towards himself, and in so doing transform the character into himself, becoming one with it. Chekhov agreed that the character is in front of the actor but he believed that the actor should move himself towards the character and transform himself into the character. The important distinction between the two approaches is that if the actor does what Stanislavski suggests, then the character will be subjected to the ego of the actor. This in the end will present some problems that the actor will have to deal with. The approach suggested by Chekhov frees the actor of many problems, because now the actor's ego is subjected to the ego of the character. It becomes easy to do what the writer requires. One way of becoming the character is to change your body into the character's body. It sounds impossible to do this, but it isn't. Already you have achieved so much with your life-body and once again we can work with this to do what we need to in order to transform. Try this: imagine that your neck is twice as long as it actually is. You cannot stretch it physically, all you need to do is manipulate the life-body neck and imagine that it is twice its normal length. As you can see it is quite easy. It is an energetic shift, yet it changes how you feel about yourself, it changes your psychology. Tell me now, how have you changed?

S: *The long neck feels comical.*

Is that all you can say about it?

S: *I feel somewhat aloof or arrogant, above everyone else.*

S: *I feel a bit lazy as if I cannot be bothered to exert myself.*

Okay, this is good because in one instant you changed your whole point of view of the world, and how you will relate to it. For you this is the psychology of the person with a long neck. Now you are beginning to understand how the psychology and the body are connected, that they are one. It's the inner body – it's the inner neck. It is the imaginary body. There is nothing physical about it, except a resonance to the inner change. If it is only physical, it will wear you out because you will become tired and tense. It is also impossible to change the physical body. Let's play with this as if it is a secret; it is gold. The problem with gold is that when you show it to people they will steal it from you. Just live with it. Just know that it is there.

'I have it, I know it will support me.' It is a concentration.

It is something coming from the life-body, and it is inspiring. Nice, isn't it? This is not you, not your normal self.

If you just say, 'It is', then that is what it is. Sometimes you just see what the character looks like, so incorporate that image and just allow it to be.

Are you getting inspiration from your body?

Great!

The image must hold you. You appear stiff when you hold the image. If the image holds you then you are free with it and it leads you and your talent follows.

Imagine that your hands are made of delicate crystal glass. Remember that they are hands, and so you must use them as hands even if you are playing with the image of them as glass. Look through your pockets. Let's play catch with this ball. How will you catch it with these glass hands? Shake hands as you greet each other. Quickly hand someone an object, touch your face, button your shirt, do as many things as you can using your hands.

Drop these images of the long neck and the glass hands. Imagine that your neck is an inch long, in fact imagine that you have no neck. How will you turn your head now? What must you do to look behind you, or to look up at the sky?

When you feel comfortable with this neck, change the hands, imagine that you have very short and stubby fingers like little sausages. Play catch and do various things with your hands.

Now imagine that your feet are very large. Allow it to be so. Do not show me that you have large feet. Just use these large feet to walk. Be discreet with it and you will find how easy it is to transform into another person. If you see it, then you can change into that. Any part of the body can change and it will change you.

We are looking for a new psychology and our way is through the body.

14 ATMOSPHERE

Today we will work on atmosphere, and we will see if we can move the space. Only Chekhov talks about atmosphere as something that can lead an actor in rehearsal and performance. Playwrights often rely on atmospheres to describe the scenes in their plays. Actors and directors often ignore these signals given by the playwright. Atmosphere is one more way we can approach the play using the How. Atmosphere means the space that is around us. The space we occupy. It is possible to manage a certain amount of space. We can work with the space that is immediately around our bodies. We are already in contact with it, as you will see.

Approach another person and stop at the point you feel socially comfortable to stand by that person.

This is something that you can feel. When you feel it, you will stop.

Go a bit beyond it now and feel what it is like to be in this social discomfort.

Return to the place where you are as close as you can be, yet you still feel comfortable.

Gently touch hands.

Leave your arm and hand where they are, and take a step or two backwards.

Your hand is now a certain distance from your body. Imagine that your hand is gently resting on the membrane of a bubble. You are inside this bubble and it surrounds you. You are standing in the middle of it.

Using your hands, define your space, your bubble. Stand there and try to become conscious of this bubble, which is also behind you. You are three-dimensional. The bubble is three-dimensional.

Walk around and be aware of your bubbles, they will come with you as you walk, and you will always be inside of them, standing in the middle of them. Bump the bubbles of the other people. Bounce off gently and move on.

Now stand within your bubble, isolated from each other.

Bring your hands up to your chest and reach out as if your hands are the antennae of a snail.

Close your eyes and reach out, encounter the imaginary membrane of your bubble. Move your hands away from you just like a snail sends out its antennae.

When you make contact with the membrane of your bubble, pull your snail antennae hands back into you like a snail would at the moment of contact, that is the moment you need to concentrate on. Make the reaction electric. It is like the surprise of touching something you don't expect will be there. In that moment of contact with the membrane, there is a *charge*. It is an instantaneous expansion, then a contraction as you pull your hands back to your chest. The movement is soft. The moment is sensitive.

Try to feel yourself in the middle of this bubble. The space within the bubble is the space we will work with.

Become aware of the fact that you are breathing. You do not need to change anything about your breathing; just be aware that you are breathing in and out. You are breathing good clean air.

Begin to move and realize that you are carrying your bubble with you everywhere you go.

Now we are going to fill the bubble with something to see how it will influence you. Imagine that the air inside your bubble is filled with the stench of urine. Because this odour is held within the membrane of your bubble, you cannot escape it.

Continue to breathe. Pay attention to what is happening to your body. How does the imagination of this smell affect your movements? How do these movements influence your relationship to the world around you?

Talk to each other, but don't talk about the urine.

See if you can laugh while you are doing this. How is your laughing affected?

Receive the 'atmosphere' within your bubble and live with it. This is how things are at this moment. Find various activities to occupy your attention.

Let the 'atmosphere' touch you.

Are you still breathing? Has your breathing changed?

We can make this go on for as long as we desire, and we can also end it when we choose to.

Allow the smell of the urine to go away now. The air is returning to good clean air. Keep breathing. Notice how your body feels now, and how your relationship to the outside world has changed because your body feels different.

The air within the bubble will now be filled with smell of lilacs.

Keep breathing. Breathe in the aroma of lilacs.

Notice what happens to your body as soon as you imagine this new scent filling your bubble. You cannot escape it. The fragrance of lilacs is in every breath you take.

Speak to each other, but do not speak about the lilacs. What is happening to your body?

Laugh. How is it to laugh now?

Find various activities to occupy your attention. Let this 'atmosphere' touch you. Keep checking in with your body.

Let this imagination go. The air will return to good clean air with no smell of lilacs.

Pay attention to your body. You changed the space surrounding it.

What has happened as a result?

Now fill the space with very thick dust.

Waving at the dust will do nothing for us. It is a psychological change we seek, not a pantomime. It is about living with this dust, not showing that it is there. Coughing is the same thing as waving at it. These activities show us that dust is in the air. You need only to *feel* it.

It is an imagination of horrible thick dust.

There is no escaping it. How is the body reacting to this?

Speak to each other, all the time you are breathing the air that is full of dust.

Laugh. Is this laughter different than the last time you laughed?

Find various activities to occupy your attention.

Let this 'atmosphere' touch you. Keep checking in with your body.

Improvise with each other. Everyone is sharing this atmosphere. It is something we have all agreed on.

Now let it go. Return to breathing good clean air. Find a chair to sit in so we can talk for a bit.

Spy back on what we just did: urine, lilacs, dust.

If it were possible that the space could move, which direction did it move? The movement is in relation to you standing in your bubble. It is an odd question. The answer requires you to use your imagination. But your memory will serve you because you were all reacting to the urine. If the space in your bubble could move what direction, was it moving when the bubble was filled with the stench of urine?

S: 1 *It was coming towards me, assaulting my nose and face.*

  2 *I agree. It was coming towards me from all directions.*

  3 *It seemed to be contracting all the time.*

Does everyone agree with this? Yes? Good!

That's great! You were able to experience the space as being in movement. This is a small amount of space. It is an easy way to begin this investigation. As I said, we already can manage this small amount of space right around us. You all agree that the space was contracting. What happened to you, to your bodies while you were responding to this space contracting?

S: 1 *It was very distracting.*
2 *I felt cut off from many things.*
3 *I was limiting my sensory input.*
4 *I felt I was trying to make myself smaller.*
5 *What I found interesting was not being able to talk about it. It was there but our improvising was not about the urine.*
6 *We all felt it and we were changed by it.*
7 *I was very uneasy with myself and with the others.*

What happened when you changed it to lilacs?

S: 1 *It definitely expanded.*
2 *Yes the space was moving away from me.*
3 *The space was moving out and away and it was taking me with it.*
4 *The space moving like this seemed to pull the laughter right out of me.*

You all agree it was expanding? Okay, what did it do to you?

S: 1 *I wanted to engage with everybody.*
2 *I really enjoyed being with people.*
3 *It was like being at a party full of people I love to be with.*
4 *I felt calm, at peace with whatever happened.*

What about the dust? What did this imagination do to the movement of the space?

S: 1 *It was pressing on me slowly, almost squeezing me.*
2 *Yes, I agree. It was also a contraction, but a different quality.*
3 *The urine was a hard kind of direct movement. This was soft and swirling.*
4 *But it definitely was contracting in relation to me.*

What was the outcome of it? What happened to you because you were in this dust filled bubble?

S: 1 *I understand the value of not talking about the dust.*
   2 *If we did, it would be boring.*
   3 *I felt as if I was involved in a conspiracy.*
   4 *The improvisation was loaded with secrecy and meetings between small groups. No one was very forthcoming.*
   5 *My eyes were really affected by the dust and I noticed that I started to dislike everyone because of this. It was like I didn't want to see them any more.*

That was a very good investigation. You were successful, so we can move on. When Chekhov spoke about the atmosphere, he was referring to something less crude than we just did. He was talking about feelings in the air, not substances. What you just explored could be useful atmospheres, but he had something different in mind. I chose to introduce atmospheres with these three, because you have already experienced them. They are a bit more tangible than 'feelings in the air'. We were able to accomplish something with that exercise. But we need to become more poetic about it and we will see what can happen.

Return to your bubbles. Define them and feel yourself within that space. You don't have to change your breathing but be aware that you are breathing. Imagine the air you are breathing becomes filled with Disaster.

S: *Has a disaster happened?*

No, please do not think about a specific disaster. Let us just say that the air you are breathing is filled with disaster. With each breath this disaster in the air goes deeper and deeper into your lungs. It is in the air; move around and breathe this disaster. If you think about a specific disaster you may be seduced into playing the disaster. If there is no real disaster, then you may be conflicted. There is a certain quality of being, of feeling in the body that accompanies a disaster. That is what we are looking for. The scene you are in could be that of a wedding, but this particular wedding is surrounded with an atmosphere of disaster.

At this point in the exploration, you should direct your attention to your bodies. What is happening to the body, because you are in this atmosphere of disaster?

If you put your attention there, you will receive useful information.

If the space could move in relation to you, the body will tell you which direction it is moving in. Let your body react to the space, it will lead you to act. But first you must react.

Did that work for you? Could you feel the direction?

S: 1 *The space was pressing down on me. I felt heavy.*
   2 *It was a very strong sensation. It was falling onto my shoulders.*
   3 *I felt a weight on my head and my face felt long.*
   4 *My breathing changed, I was very aware of how I was breathing. That was interesting, because I am not usually aware of how I breathe.*
   5 *The space was definitely moving downwards.*

We have found something very important, something that will allow us to recreate this atmosphere again and again. We seem to have a consensus that the space felt as if it were moving down onto the head and shoulders.

You already have had inner sensations of falling and you know what that is.

Now it is the space outside of us that is falling. Because the 'falling' space is outside the body, it touches the body and the body reacts.

It is so wonderful to hear how your responses have become physical, and that you have already developed a new relationship with your bodies. You are able to hear it speaking to you, and you can use that to act with.

15 *DESIRE UNDER THE ELMS*

Let's see if we can take the word 'disaster' out of the equation.

Usually actors lose the atmosphere when they begin speaking, because they get involved with the words. We don't want to play a disaster, we want to be affected by the atmosphere of disaster.

In the first scene we are working on, Cabot returns to the farm with his new bride. For Eben it is not a good situation. It is a kind of disaster for him, yet the scene is actually a homecoming. Cabot reasserts his authority, and offers a real threat to Eben, and the plans Eben has already put in motion. It is not a positive thing for Eben. Most negative things are downward moving.

Now please don't worry about a disaster. Try to engage the space and experience the downward movement – so you don't have a disaster to think about, just work with the space moving down upon your shoulders.

For Abby she has anticipated this moment of meeting Eben, and this atmosphere is not what she expected. She has to react to this, it is in the air. The whole event is a surprise to Eben, it came out of nowhere and he is reeling in this atmosphere. They have two completely different points of view, yet they are in the same place and the same air surrounds the two of them.

Let's get the space moving and then we can take it from there.

Something is happening! I can feel the atmosphere you have created. That is one of the greatest benefits of the atmosphere, how it reaches past the stage into the audience. We cannot expect the audience to act our parts for us, but we can give them something that they can feel. That is the atmosphere. I am your audience now and I can feel this atmosphere. Can you feel it?

Good. Stop doing it now. Stop moving the space and we will feel it go away. That was a very good first step towards playing with the atmosphere. How did it feel?

S: 1 *Because you took away the disaster, and we were coming to it with this scene, it felt different than the disaster. Not so strong and specific as a disaster, but it was still quite uncomfortable and unsettling.*

2 *I think it would have worked for the scene. I was disappointed we had to let it go. I was ready to act the scene. Abby was present for me. Her arrival*

into this household was clarified, how much work it would be, the difficulties she would have to overcome. I could feel all that.

3 I could feel how Eben was very upset and didn't know what to do or how to respond. It wasn't a disaster at all, but it was not a good thing, and I felt that pressing on me.

Good. As I said, I could feel it too. Let's see if you can create the atmosphere of a downward moving space again. And then just begin to touch the words easily. Play around, but put more attention on the atmosphere than on the script. Allow yourselves to react to the atmosphere and see how the words will come out of you.

Stop. You are giving too much attention to the words, and you are letting the atmosphere slip away from you. Stay in the atmosphere. Just sit down or get up in harmony with the atmosphere, and you will receive many things from it. When you are confident that you can maintain this atmosphere, then try speaking. I will continue to stop you if it slips away from you. This is what we are working on tonight. I know that you know the script, so I am not interested in that right now. Let the atmosphere be everything. In this way we can develop a way to work, and what our specific point of focus should be. So try to create this atmosphere again, and see what you will find in it. It is a new way to rehearse so it will take some getting used to. Soon you will always look for atmospheres, because you will believe in the power of what is around you. Do it again, please.

Good, the improvisation went well this time, because you were clear about what you were focused on. How did it feel for you?

S: 1 As Eben, I wanted to run away from them and the scene, but I had to stay and deal with it. That was very powerful. It gave my words a mysterious force. But I could accept it as the actor in the middle of it all.

2 *I agree with that. As Eben, I felt that the only way to fix these difficult feelings was to throw the both of them out, and my words were saying this, even if they were not saying it.*

3 *As Abby, I felt so isolated. I felt a deep resentment for Cabot, but I did feel sorry for Eben, who seemed so far away from me.*

Let's try the scene with the atmosphere and the text. We will do it with one Abby and one Eben. Remember it is only a rehearsal. Keep your focus on the atmosphere and your attention on each other. Just let the words follow everything else.

That was good. Those of you watching, what did you see or feel?

S: 1 *I felt that both of these characters were definitely in the same place surrounded by the same thing. It seemed to hold them together, even though they were very far from each other. I mean their desires were clearly different from each other but something was unifying them.*

2 *Their actions were very clear. They moved as if in a kind of spell. It was rich.*

Did you think it was correct? Was it the right atmosphere? Do you think it could have been different with a different atmosphere?

S: *It was the atmosphere that you chose for them. I figured it must be correct if that is what you told them to do. Could there be another atmosphere for the scene? Something exciting was happening between them. But maybe it was not the right atmosphere.*

It was only a rehearsal. We were trying something. It was something that they could do, because they had already developed it. So I asked them to try it. Chekhov said, 'Rehearsals are for finding better things.' Yes, there could be other atmospheres to explore. In the end we would choose the one that best suits the needs of the scene.

What would it be like if we looked at the atmosphere of Suspicion, and then did the scene? Fill your bubble with suspicion and then breathe it into your lungs. Just like you did with disaster. Please do not feel obliged to become suspicious. That is not the atmosphere. In fact, that will kill the atmosphere. Everybody try this. We are trying to find how the space moves when the atmosphere is full of suspicion. Breathe it in. Let the atmosphere act you.

Okay, that was very powerful, I am sure you will agree. What can you say about the direction the space was moving in?

S: 1 I felt it was coming at me in a kind of point, right at my forehead.
   2 Yes, it was moving towards me like a V, and the point of that V was my face.
   3 This is incredible, I also felt like it was touching my face, but it was not a V. It was kind of spiralling toward me.

All right, this is good. Let's drop the word suspicion, and work only with the movement. Then you will not be tempted to act suspicious. Instead you will react to the movement of the space. Get it going, and begin an improvisation. I think you understand now what to focus on. Go ahead and play with it, be played by it.

S: I think this was better. It was a stronger clearer and easier thing for me to connect with.

Perhaps you are becoming more familiar with how to do it.

S: No. I think it was more correct for the scene. The kind of discomfort was much more active. I was not working with suspicion. I was reacting to the specific direction and movement of the space; it seemed more to the point of the scene.

Why don't you do the scene now? Can we have an Abby join him? Get the space moving. When you feel the atmosphere is there, then begin to speak the lines and see if what you suspected is true.

That was good. It must have felt good because I felt something from it. How was it for you?

S: 1 *It was good. It was not what I expected, but I felt something more because I was living with this movement that would not stop, yet I was able to act with her.*

2 *Yes, it was a powerful experience, but it was surprisingly easy to maintain. Did we maintain it?*

I think you did. What you said about the ease of it, I know this is true. Once you give yourself over to the possibility that it can happen, and you set the space in motion, it will continue for as long as you want it to. That was good work. I agree with you, it is more appropriate to this scene. It is full of not understanding what is happening, not quite believing that all of this is really happening. And that is true for the both of them.

With time and practice we can control more and more space. It is a matter of developing the concentration. But what you accomplished tonight was successful, because you all felt it. You really contacted something that was intangible, and that is the real power in the technique.

6

MASTERY OF THE TECHNIQUE

There is nothing personal about this approach to acting, yet it opens to us our true selves. When we first come in contact with the technique, we have two minds about it. It appeals to us because it promises so much, yet we have doubts about it, because we do not understand how something like movement can bring about anything organic and truthful. The technique is often described as 'outside in', while other approaches, involving more thinking and less movement, are described as 'inside out'. Neither of these descriptions correctly portrays their approach, nor do these descriptions say anything useful about the process of acting. They are used as value indicators. Because of this, there are definite camps about which approach has more value.

The body is outside in relation to the imagination, but it is the imagination that must come out to find the body. In my opinion, the Chekhov technique is not 'outside in'; it is perhaps the most 'inside out' process there is. The technique promises access to inspired acting. The steps to inspiration are:

1   Imagination (inside).
2   Concentration (inside to outside).
3   Incorporation (outside to inside).
4   Radiation (inside to outside).
5   Inspiration (outside to inside to outside).

In an ideal world there is no gap between the imagination and the means to express it, but in reality, a gulf separates them. The life of the artist spans that gulf. For the actor, it is the body that has to be mastered; there is no other way, because that is the instrument.

We have to know what is possible for us. We have to learn how to manage our time and put our priorities in perspective. We have to work diligently to soften the body so it can serve the imagination. The body is hard, real life takes the suppleness away from us, and the body does not always follow so easily the commands we give it, especially when the commands demand performance. If we are going to befriend the physical body, which is initially the enemy of the actor, then it will take time. By exercising consciously and correctly, we find a way to soften the body. We cannot do anything with our hard bodies. We cannot incorporate an image. We cannot express our feelings or intentions. The learning begins with discovering that the body is the instrument. Then we have to begin as beginners, as if we didn't know we had a body, and we are just now discovering it. This little shift in perception is necessary. When we have a new realization of what the body is, then we can develop it and work it in the right way to serve us.

The world is vast, and our little lives are so limited. Once we realize that we can be led by the imagination, then we will develop it because we will see that it is a power. When the imagination is developed, we can rest assured that everything we need as actors can be found within. We will find a clear means to express ourselves consistently in performance. We must be willing to reach beyond our normal sense of self into a new world of archetypes and images. Making contact with a script in an archetypal way releases the powers that are living within the imagination of the artist.

Deirdre Hurst du Prey told us what Chekhov demanded of his students:

> You must develop a moral responsibility towards the things you present to the world. You need to know the effect you will have upon the audience, and how it will reach them. You have to know that you are there to make a generous contact with the audience. They want this from you, and you must be willing to oblige them.

You are not acting for yourselves. It feels good to be an actor acting. But it is larger than that, it is more important than just doing it for yourself.

Mastering this technique means embracing new ideas about being a human being. It can't happen otherwise. The activity of searching for moral responsibility is an indirect appeal towards finding our connection to what Chekhov called the 'creative individuality' of the artist.

Claiming that part of ourselves is the work we need to set in motion. Everyone knows that it takes time to master anything. Making contact with the creative individuality will take time, changing the body will take time.

What the principles and the tools offer us are delights, because they come to us so freely. But these principles and tools are in many ways merely tricks that help us create the magic of acting. The tricks will appear as many disjointed parts until we find a way to unify them within us. The creative individuality can do that for us, because it teaches us a way to use these tools, as we need them.

In my classes, I try to teach the technique in the most practical way possible, so it can be used. Many actors come to me for advice on how to utilize the technique after they have learned some parts of it. They tell me they understand the technique, but not exactly what parts to use or when to use it. These questions troubled me, because the answers to their questions seemed so obvious. I wondered why they could not determine these things for themselves. In class they could do the exercises well. They were able to speak about the value of the work, and about what they were experiencing. Why could they not put it into play for themselves? I reflected on my own training and development, then I realized what I had forgotten. I too had struggled for a long time with the same problems and questions. I had forgotten that it took me ten years to 'get it'. As a teacher, I had the erroneous idea that an artist could master this technique in two years. This is true for only two parts of the learning, the *What and the How*. To be complete, the *Who* must be developed as well. In this case the who is you, the artist.

While developing the physical body, the creative individuality becomes engaged. If you are not attentive, it is possible to miss this connection and then the training can become one-sided and even take more time to master. You will be able to physically do it, but you will not be free in it. You can be led by a director or teacher and achieve some wonderful things, but you will not be in control. This feeling of being in control of all that you are doing is special and belongs to accomplished artists.

After you know how you feel about the world you live in, then you will have the power to make artistic connections. Knowing *how* you feel is found in your daily moment-to-moment reality. Knowing *where* you feel things, and *how* you feel things, gives you the deftness to play with the technique. The creative individuality is that function within which brings the real world into a relationship with the imaginary world. In striving to make the imaginary world appear real, actors must set things within the body. Without the awareness of how the sensations affect the body, we are not in a creative relationship with our world.

At the Michael Chekhov Acting Studio in New York City, a course of study is offered based on the five guiding principles. The approach is split into three focuses. The first is called 'I'. This is work on the actor's body. The basic principles and tools are stressed. This focus is for the actor to discover what feeling and moving the body can do. There is a very quick sense of accomplishment here, because the body is so responsive to the exercises. The next focus is called 'WE'. Here the actor is brought into relationship with the space he occupies, a relationship with his fellow actors, and a relationship with the imaginary world of the play. It is an expansive investigation, because the attention is outside of the body, but the impressions are felt and expressed by the body. The third focus is called 'THEY'. It is connected to the performance, to the audience, and to the world beyond the footlights. Here the actor becomes connected to the affect he is having upon the world and how he can express his true self. It is in each of these focuses that the creative individuality is stimulated and encouraged to play a part in his artistic understanding. The creative individuality unites these three focuses and gives the actor a *Feeling of the Whole*. This feeling of the whole is one of the Four Brothers, one of the dynamic principles that will always keep us on the mark. When we act with the feeling of the whole, we are working as artists, and we can feel confident that we have embraced the technique and we are well on the way to mastering it.